THE MATHEMATICS
OF RESERVOIR SIMULATION

FRONTIERS IN APPLIED MATHEMATICS

H. T. Banks, *Managing Editor*

This series is intended to serve as a provocative intellectual forum on both emerging or rapidly developing research areas and fields that are already of great interest to a broad spectrum of the scientific community.

The Mathematics of Reservoir Simulation is the first volume in this series.

Editorial Board Victor Barcilon   Richard E. Ewing
                   John D. Buckmaster Kenneth I. Gross
                   Robert Burridge   Charles J. Holland

THE MATHEMATICS OF RESERVOIR SIMULATION

RICHARD E. EWING, EDITOR

PHILADELPHIA
1983

Copyright © 1983 by Society for Industrial and Applied Mathematics.
All rights reserved.

Library of Congress Catalog Card Number: 83-51501
ISBN: 0-89871-192-4

CONTENTS

Contributors vii

Foreword .... *H. T. Banks* ix

Preface .... *Richard E. Ewing* xi

Chapter I.   Problems Arising in the Modeling of Processes for Hydrocarbon Recovery .... *Richard E. Ewing*   3

Chapter II. Finite Element and Finite Difference Methods for Continuous Flows in Porous Media .... *Thomas F. Russell and Mary Fanett Wheeler* 35

Chapter III. A Front Tracking Reservoir Simulator, Five-Spot Validation Studies and the Water Coning Problem .... *James Glimm, Brent Lindquist, Oliver McBryan, and L Padmanabhan* 107

Chapter IV. Statistical Fluid Dynamics: The Influence of Geometry on Surface Instabilities .... *James Glimm, Eli Isaacson, Brent Lindquist, Oliver McBryan, and Sara Yaniv* 137

Chapter V.   Some Numerical Methods for Discontinuous Flows in Porous Media ...... *Phillip Colella, Paul Concus, and James Sethian* 161

CONTRIBUTORS

PHILLIP COLELLA received his Ph.D. in Applied Mathematics from the University of California, Berkeley in 1979. Since then, he has been at the Lawrence Berkeley Laboratory, where he is a staff scientist engaged in research in numerical methods for fluid dynamics.

PAUL CONCUS received his Ph.D. in Applied Mathematics from Harvard University in 1959. Since that time he has been principally at the Lawrence Berkeley Laboratory, where he is a staff senior scientist. He also is on the faculty of the University of California, Berkeley.

RICHARD E. EWING, appointed to the University of Wyoming as the J. E. Warren Distinguished Professor of Energy and Environment, holds a joint position in the Departments of Mathematics and Petroleum Engineering. He came to the University from a position as coordinator of mathematical analysis for Mobil Research and Development Corporation in Dallas. After receiving his Ph.D. under the direction of Professor John Cannon at the University of Texas at Austin in 1974, Dr. Ewing held academic positions at Oakland University and the Ohio State University and visiting positions at the University of Chicago and the Mathematics Research Center in Madison, Wisconsin. His principal research interests are in applied mathematics, in numerical analysis and solution of partial differential equations, and in numerical reservoir simulation.

JAMES GLIMM is a Professor of Mathematics at New York University. He received his Ph.D. from Columbia University in 1959. He was previously on the faculties of the Massachusetts Institute of Technology and Rockefeller University. He has worked in the areas of functional analysis, partial differential equations and mathematical physics prior to his involvement in computational fluid dynamics.

ELI ISAACSON received his Ph.D. from New York University in 1979. He joined the Mathematics Department of the University of Wyoming as Assistant Professor after a postdoctoral position at Rockefeller University. His work is in the areas of nonlinear differential equations, analysis and applied mathematics.

BRENT LINDQUIST has a postdoctoral appointment at the Courant Institute, New York University. He received his Ph.D. from Cornell University in 1981 in Theoretical Physics. He has done work in the area of elementary particles and fields and currently in computational fluid dynamics.

OLIVER MCBRYAN received his Ph.D. from Harvard University in 1973 in Theoretical Physics. After postdoctoral positions at the University of Toronto and Rockefeller University, he taught at Cornell University. Since 1979 he has been a professor at the Courant Institute, New York University.

L PADMANABHAN holds a Ph.D. from Northwestern University in the field of Chemical Engineering. For the last ten years he has been with Chevron Oil Field Research Company, La Habra, California, where he is engaged in research in mathematical modeling of multiphase fluid flow in porous media. He is currently concerned with the development of numerical methods for solving parabolic and nonlinear hyperbolic PDEs that arise in the simulation of enhanced oil recovery processes. His other research interests include algorithms in computer graphics and inverse problems in history matching. He has published over twenty papers in diverse fields such as chemical reactor stability, control and optimization, numerical solution of stiff ODEs, stochastic filtering and inverse problems in well testing.

THOMAS F. RUSSELL is a research mathematician with Marathon Oil Company in Littleton, Colorado. He received his Ph.D. in Mathematics from the University of Chicago in 1980, under the direction of Jim Douglas, Jr. He is interested in theoretical and computational aspects of petroleum reservoir simulation.

JAMES SETHIAN received his Ph.D. in Applied Mathematics from the University of California, Berkeley in 1982. He is currently an NSF Postdoctoral Fellow at the Lawrence Berkeley Laboratory.

MARY FANETT WHEELER is a Professor in the Mathematical Sciences Department at Rice University, Houston, Texas and is a member of the SIAM Council. She received her Ph.D. in Mathematics at Rice University in 1971. Her major research interests are the numerical solution of partial differential equations and the numerical simulation of petroleum reservoir engineering problems.

SARA YANIV has a postdoctoral position at the Courant Institute, New York University. She received her Ph.D. from Tel-Aviv University in 1979 in Applied Mathematics.

FOREWORD

This is the inaugural volume in the SIAM series *Frontiers in Applied Mathematics*. This continuing series will focus on "hot topics" in applied mathematics, and will consist of, in general, unrelated volumes, each dealing with a particular research topic that should be of significant interest to a spectrum of members of the scientific community. Distinguished scientists and applied mathematicians will be solicited to contribute their points of view on "state-of-the-art" developments in the topics addressed.

The volumes are intended to provide provocative intellectual forums on emerging or rapidly developing fields of research as well as be of value in the general education of the scientific community on current topics. In view of this latter goal, the solicited articles will be designed to give the nonexpert, nonspecialist some appreciation of the goals, problems, difficulties, possible approaches and tools, and controversial aspects, if any, of current efforts in an area of importance to scientists of varied persuasions.

Each volume will begin with a tutorial article in which technical terms, jargon, etc. are introduced and explained. This will be followed by a number of research-oriented summary contributions on topics relevant to the subject of the volume. We hope that the presentations will give mathematicians and nonmathematicians alike some understanding of the important role mathematics is playing, or perhaps might play, in what academicians often euphemistically call "the real world." We therefore expect each volume to contribute some further understanding of the important scientific interfaces that are present in many applied problems, especially those found in industrial endeavors.

At the printing of this first volume, a number of other volumes are already in progress. Volume 2 will focus on seismic exploration, Volume 3 on combustion, and Volume 4 on emerging opportunities related to parallel computing. Other topics in scientific computing are among those currently under consideration for future volumes. Members of a rotating editorial board will encourage active participation of the mathematical and scientific community in selection of topics for future volumes.

FOREWORD

The series is being launched with several volumes in "energy mathematics" and the editorial board gratefully acknowledges the faculty and administration of the University of Wyoming for their cooperation in connection with the Special Year on Energy-Related Mathematics held in Laramie during the 1982–83 academic year. A great deal of the planning of these first volumes was facilitated through contact with visitors to the special program at the University of Wyoming.

H. T. BANKS

PREFACE

Over the past few years, the interest in the numerical modeling of fluid displacement processes in porous media has been rising rapidly. The emergence of complex enhanced recovery procedures in the field of hydrocarbon extraction techniques has emphasized the need for sophisticated mathematical tools, capable of modeling intricate chemical and physical phenomena and sharply changing fluid interfaces. The goals of this volume are to explain to the mathematical and scientific community which problems need to be addressed, why they are difficult, what has been done previously to treat these difficulties, and which new techniques appear to possess potential for obtaining good simulation results.

The first chapter presents an introduction to the physics of a wide variety of fluid displacement techniques, illustrating the complexity of the associated processes and difficulties inherent in the modeling process. Using certain basic physical "laws," model partial differential equations are derived which are typical of the mathematical models used to describe the flow of fluids in a porous medium. The variety of physical phenomena discussed illustrate that many different mathematical tools are necessary to model the complete spectrum of enhanced recovery procedures. Special mathematical problems which must be addressed, such as modeling local behavior around the wells with point sources and sinks and coning studies, modeling dispersive mixing and viscous fingering, elimination of grid orientation and numerical dispersion problems, and modeling of sharp moving fronts with concern for accuracy and stability, are presented. Finally, the difficulties of applying the mathematical techniques on very coarse computational grids mandated by the enormous size of standard reservoir simulation problems are addressed. A literature survey indicating articles for further reading is also presented.

In Chapter II an argument is developed, with extensive references to the petroleum literature, for the need to incorporate modeling of physical dispersion phenomena in most of the enhanced recovery procedures. The dispersion tensor is derived for the miscible displacement problem and the difficulties in the associated convection-diffusion type of numerical problems are described. Then a discussion of the existing theoretical and computational literature in this area as well as perceived directions for future research are presented.

In Chapters III–V, a different approach to reservoir simulation is presented. These chapters discuss a variety of techniques and applications of front tracking. Under the explicit assumption that capillary effects, dispersive mixing, and physical phenomena around the fluid interfaces are either unimportant or incapable of being resolved on a coarse-grid level, the convection-diffusion equations developed in Chapters I and II are reduced to systems of nonlinear hyperbolic conservation laws.

Chapters III and IV utilize front tracking methods to address the modeling of two important types of fingering phenomena. Due to gravitational segregation, oil is often found above a layer of water. Since water flows more readily in the reservoir, if a well is open to flow near this oil-water interface, the water will cone up into the well, reducing or possibly stopping the oil production. In Chapter III, front tracking techniques are applied to model this water coning problem.

An important phenomenon which drastically reduces hydrocarbon recovery is the bypassing of oil by injected fluid due to the growth of long narrow fingers caused by the instability of displacing a viscous fluid by a less viscous one. Although the instability is initiated on a microscopic level and cannot be modeled in that regime, the effect should be incorporated in some statistical manner in a global model. Aspects of this modeling procedure are discussed in Chapter IV. Finally, in Chapter V, a variety of methods for tracking discontinuities for hyperbolic partial differential equations are presented. These techniques include random choice methods, a higher dimensional analogue based on "volume of fluid" constructions, higher order Godunov schemes, and various operator splitting techniques. Although shock tracking techniques cannot, by themselves, resolve the important physics of the enhanced recovery processes in the regions of fluid interfaces, it is hoped that these methods can be coupled with other, local techniques, to simulate the more complex secondary and tertiary recovery methods.

R. E. EWING

THE MATHEMATICS
OF RESERVOIR SIMULATION

CHAPTER I

Problems Arising in the Modeling of Processes for Hydrocarbon Recovery

RICHARD E. EWING

1. Introduction. The objective of reservoir simulation is to understand the complex chemical, physical, and fluid flow processes occurring in a petroleum reservoir sufficiently well to be able to optimize the recovery of hydrocarbon. To do this, one must be able to predict the performance of the reservoir under various exploitation schemes. In order to predict reservoir performance, a series of models of reservoir processes are constructed which yield information about the complex phenomena accompanying different recovery methods. In this volume, various approaches to the use of mathematical models for reservoir simulation problems are presented and compared.

There are four major stages to the modeling process for reservoir simulation. First, a physical model of the flow processes is developed incorporating as much physics as is deemed necessary to describe the essential phenomena. Second, a mathematical formulation of the physical model is obtained, usually involving coupled systems of nonlinear partial differential equations. Third, once the properties of the mathematical model, such as existence, uniqueness, and regularity of the solution, are sufficiently well understood and the properties seem compatible with the physical model, a discretized numerical model of the mathematical equations is produced. A numerical model is determined which has the required properties of accuracy and stability and which produces solutions representing the basic physical features as well as possible without introducing spurious phenomena associated with the specific numerical scheme. Finally, a computer program capable of efficiently performing the necessary computations for the numerical model is sought. The total modeling process encompasses aspects of each of these four intermediate steps. Important problems in numerical reservoir simulation arising at each step of the modeling process are described later.

Finally, the modeling process is not complete with one pass through these four steps. Once a computer program has been developed which gives concrete quantitative results for the total model, this output should be compared with measured observations of the physical process. If the results do not compare extremely well, one should iterate back through the complete modeling process, changing the various intermediate models in ways to obtain better correlation between the physical measurements and the computational results. Usually many iterations of this modeling loop are necessary to obtain reasonable models for the highly complex physical phenomena described in this volume.

The aims of this chapter are to present an introduction to the physical properties of a variety of enhanced recovery techniques, to discuss a series of models needed to describe these processes, to indicate some of the important problems encountered in building these models, and to point out some of the ways newer mathematical tools have been successful in addressing these problems. As the physical phenomena being modeled become more complex, the engineers and physicists are realizing the inadequacies of standard modeling tools and are asking the mathematicians for increasingly more help. The mathematical community has already contributed significantly to the aspects of linearization and linear solution of models for reservoir flow. The increasing use of large coupled systems of nonlinear partial differential equations to describe the movement of sharp fluid interfaces is pointing out very difficult problems in the theoretical aspects of the partial differential equations, the numerical analysis of various discretization schemes, the development of new, accurate numerical models, and the computational efficiency of the resulting discrete systems which must be solved for practical modeling of enhanced recovery processes. It is hoped that this volume will help to introduce the mathematical community to the wealth of interesting and difficult mathematical problems arising in the area of reservoir simulation and to encourage its participation in this important area of scientific research.

2. Description of problems in recovery simulation. In order to understand the complexities of the development of a physical model in the context of reservoir simulation, a brief description of various chemical and physical phenomena accompanying fluid flow in porous media is first presented. A common misconception about hydrocarbon recovery is that oil or gas is found in large pools in underground caverns and must be pumped out of the reservoir in a manner similar to pumping a liquid out of a storage tank. This is *not* the case. In general, the hydrocarbon is trapped in the microscopic pores of a rock, like sandstone, and will flow through the rock only under the influence of extremely large pressure differentials. The pores are formed as spaces between the sand crystals that have been compacted and fused together with various clays. A large percentage of the pores are connected and the fluids can flow through these linked pore paths. However, the paths are very small and highly irregular and twisted. Thus, instead of smooth laminar flow, the tortuousness of the paths yields an unsteady, twisting fluid flow on the microscopic level. The ratio of the volume of these pore paths available to flow to the total volume of the rock sample is termed the *porosity* of the rock and is extremely small (typically 1–20%). The type of reservoir rock present greatly influences the ability of the hydrocarbon to flow through the reservoir. Also, as the geology in the reservoir changes, the reservoir often has areas of high flow and areas of reduced flow. This heterogeneous nature of most reservoirs greatly complicates the mathematical modeling process, especially since the reservoir is inaccessible and first-hand information about its porosity can only be obtained through the wells, which may be hundreds of yards apart. See Fig. 1.

In many instances, the pressures found in reservoirs are extremely high. Often these pressures are high enough to force the resident fluids to flow through the porous medium and out of the wells without much pumping effort at the wells. This type of recovery, termed *primary* recovery, has comprised a large percentage of past extraction techniques. The simulation of primary recovery techniques and certain generalizations often termed "black-oil" models have been discussed by several authors [5], [6], [8], [27], [72], [85], [86], [88], [120]. As the reservoir pressure is depleted through fluid extraction, the hydrocarbons flow more slowly and production rates decrease. Often primary recovery leaves 70–85% or more of the hydrocarbons in the reservoir.

Subsequently, a fluid like water may be injected into some wells in the reservoir while petroleum is produced through others. This serves the dual purpose of maintaining high reservoir pressures and flow rates and also of flooding the porous medium to physically displace some of the oil and push it toward the production wells. This type of pressure maintenance and waterflooding is usually termed *secondary* recovery. In this process the water does not mix with the oil due to surface tension effects and the process is termed *immiscible displacement* [11]–[15], [36], [40], [41], [44], [48], [70], [101], [111], [115].

Unfortunately, waterflooding is still not extremely effective and significant amounts of hydrocarbon often remain in the reservoir (50% or more). Due to strong surface tension effects large amounts of oil are trapped in small pores with narrow throats and cannot be washed out with standard waterflooding techniques. Also the process of pushing a heavy viscous oil through a porous medium with a lighter, less viscous fluid like water is a very unstable process. If the flow

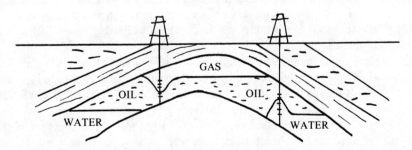

FIG. 1. *Coning of gas and water.*

rate is sufficiently high, the interface between the resident petroleum and the invading water becomes unstable and tends to form long fingers which grow in length toward the production wells, bypassing much of the hydrocarbons. Once a path consisting of water has extended from an injection well to a production well, that production well will henceforth produce primarily water which flows much more easily due to its lower viscosity. The production of petroleum from that well is then greatly reduced if not essentially stopped. This phenomenon, termed viscous fingering, is well known [16], [18], [20], [34], [63], [65], [73], [80], [81], [87], [93], [94] and is a serious problem in hydrocarbon recovery. This problem and different techniques for modeling its effects are discussed in some detail in later sections and in Chapter IV of this volume.

In order to recover more of the residual hydrocarbons, several enhanced recovery techniques involving complex chemical and thermal effects have been developed. These techniques form part of the variety of methods termed *tertiary* recovery; they are extremely difficult to model since the physical processes and accompanying phase changes must also be described mathematically. There are many quite different forms of tertiary enhanced oil recovery (EOR) techniques.

Since much oil is trapped in the pores of the reservoir because of surface tension effects, several EOR techniques involve the lowering of the surface tension to allow the oil to flow from the small pores. One method for lowering surface tension involves the use of surfactants or detergents to interact with the trapped petroleum and essentially "wash" the reservoir, similar to common uses of detergents. The chemistry and physics involved in this process includes ionic exchange and phase changes and is quite sensitive to the reservoir rock types, the salinity of the connate water, and the concentration of the surfactant. These properties must be continually determined during the simulation to yield the correct process description. Since surfactants are relatively expensive, a small slug of surfactant is usually followed by a polymer slug and then a water drive in these chemical flooding processes. The concentration of surfactant in the mixture is crucial in the simulation since it governs the physical reactions. The physical, mathematical, and numerical models for simulation of this process are continually being upgraded as the phenomenon becomes better understood. In particular, the simulation must accurately represent the diffusion and dispersion [75], [92] of the surfactant into the other fluids without allowing numerical effects which smear the interface.

Another process for EOR is the injection of gases like CO_2 or other chemical species which mix with the resident hydrocarbon with a phase change to form one fluid phase. If complete mixing or "miscibility" is attained, the fluids will flow together in one phase, eliminating distinction between phases, and complete hydrocarbon recovery is theoretically possible. Very specific percentages of the various injected and resident fluids must be present together with certain pressures for the phase changes to occur or for miscibility to be attained; thus, very accurate modeling of physical dispersion is again vital. Displacement procedures relying on this type of mixing or phase change are termed "miscible" displacements. There is extensive literature on physical [5], [65], [74], [82], [88], [94], [108], mathematical [7], [34], [46], [55], [73], [90], and numerical [12], [31], [32], [35], [37], [42], [43], [45], [46], [49], [51], [54], [55], [56], [59], [107], [109], [116] models for simple miscible displacement processes. Chapter II of this volume concentrates on surveying various miscible displacement models. In particular, the coupled system of partial differential equations for modeling the simplest miscible displacement process is developed in Chapter II.

Since the viscosities of various surfactants with water drives and fluids used in the miscible techniques are, in general, much lower than that of the resident hydrocarbon in the reservoir, the viscous fingering phenomenon described earlier is also an important problem in tertiary recovery techniques. Various models have been developed to predict the onset of fingering [16], [34], [65], [80], [81], [87], [94], to predict the finger growth process [93], [106], and to understand the statistical effects sufficiently well to incorporate aspects of fingering in simulators [20], [61], [64], [73], [113]. Chapter IV of this volume elaborates upon this last goal. Since the viscous fingering phenomenon degrades many secondary and tertiary processes so severely, much effort has been expended upon attempts to control the problem.

In one technique, polymers are added to the invading fluid to thicken or raise the viscosity of that fluid, reduce the viscosity ratio between the fluids, and control the instabilities causing the fingering. Again, the polymers are expensive and their saturation is, in general, graded from heavy at the resident fluid interface to zero in the water drive. This graded saturation helps to control the mobility of the total fluid and must be numerically monitored throughout the simulation of the displacement process to accurately predict hydrocarbon recovery. Another very important goal in the use of polymers in flooding processes is to alter the permeability or ability of the reservoir rock to allow flow in certain ways. Since the polymers are highly viscous, they can be used to selectively block certain pores or flow regions to direct the flow in a manner to optimize the hydrocarbon recovery. Research is continuing in the use and modeling of flow control via polymer injection.

Often the petroleum found in the reservoir is so highly viscous that it will not flow well or at all through the porous medium, even under extremely high pressure gradients. In this case, something must be introduced to lower the viscosity to a sufficiently low level for reasonable flow to occur. When a gas like CO_2 is used in the miscible displacement process, the phase change which achieves miscibility has the effect of producing a phase with viscosity level intermediate between those of the gas and the oil. The miscible phase flows more readily than the oil.

Another technique for lowering the viscosity of heavy oils involves the addition of very high temperatures to effectively "melt" the viscous hydrocarbons, allowing them to flow more readily through the porous medium. Getting the high temperatures down into the reservoir is not easy. A primary technique for thermal flooding is to inject either very hot water or steam into the reservoir via certain injection wells. As the heated fluids come into contact with the cooler reservoir rock and hydrocarbons, much heat is lost and the effectiveness of the process is reduced. Since the heat is the dominant physical element for the process, the thermal distribution must be accurately simulated to model the physical reactions driven by the heat. After sufficient heat is lost to the reservoir or through reactions, the injected fluid becomes too cool to be effective, the chemical reactions cease, and the process becomes ineffective. The heated fluid is injected into the bottom of the reservoir due to a gravity override problem with flow of hot fluids through the reservoir. This tendency of hot fluids to preferentially travel toward the top of the reservoir in the flow process is a phenomenon similar in the large scale to the viscous fingering process and also causes similar difficulties during simulation. The use of thermal processes is rising and the need for more accurate simulation of these techniques is crucial. Some techniques for thermal simulation appear [30], [97] and will be discussed in Chapter II.

Finally, there are massive amounts of hydrocarbon, located deep in the earth in the form of coal, tar sands, and oil shale, that cannot be made to flow in their present state under any circumstances. When these reserves are too deep to be mined, techniques for *in situ* transformation of the hydrocarbons into states which can be pumped to the surface via production wells are required. Injection and production wells are drilled into the shale layer or coal seam together with certain communication lines between the wells. The solid hydrocarbon is ignited in the ground and oxygen or air is pumped into the injection wells to maintain the *in situ* combustion. The hot hydrocarbon gases produced in the combustion process are then pumped to the surface via the production wells for use as low-grade hydrocarbon. The complex chemical and thermal processes coupled with accompanying cavity growth and overburden collapse make the mathematical simulation of this hydrocarbon recovery process a formidable task. Various mathematical models for this process involve a coupled system of up to twenty nonlinear partial differential equations [30], [68], [114]. As the process is better understood and more powerful numerical techniques are developed, more physics can be incorporated and the models will become even more complicated. At this time even simple physical models completely tax the limits of our present numerical and computational abilities. For a more complete overview and indications of the state of the art of reservoir simulation from an engineering standpoint, the reader should consult review articles by Odeh [86] and Coats [22] and texts by Aziz and Settari [5] and Peaceman [88].

Although the various types of recovery techniques described above involve many very different physical processes, there are certain common mathematical problems which must be addressed. First, the EOR displacement processes are dominated by convective flow from the injection wells to the production wells. Therefore, mathematical models for each of these enhanced recovery techniques must have strong transport terms and must possess many of the properties of first order hyperbolic partial differential equations. Another important common phenomenon in each of the models is the importance of the location of the interface between the injected fluid and the resident fluids. The location of this interface indicates how much of and where the hydrocarbons are left in the reservoir as a function of time. Knowledge of the location of the front is crucial for determining infield drilling and new production strategies to try to optimize the hydrocarbon recovery. For this reason several front-tracking techniques have been suggested for various EOR models in the literature [1]–[4], [11], [17], [26], [28], [29], [60], [62], [66], [69], [76]–[79]. Certain techniques of this type are discussed in Chapters III and V of this volume.

The location of the front is not the only important information desired at the fluid interface. Most of the chemical and physical processes which drive the EOR methods occur where the injected fluids come into contact with the reservoir fluids. The concentration of the injected fluid around the interface determines the rates and effectiveness of the chemical and physical processes and must be carefully simulated. If these concentrations are predicted inaccurately due to problems with the modeling process, the important physical phenomena will not be simulated sufficiently well to understand the crucial aspects of the EOR method. Thus, in many cases, the sole use of front tracking methods is inadequate, and techniques which give a more detailed description of the solution in the neighborhood of the fluid interface are mandatory. Furthermore, numerical diffusion and dispersion, which are common problems in modeling of convection-dominated flows must be avoided. Several methods of this type are discussed in Chapter II of this volume. One technique which may be crucial for accurate simulation of this local behavior is the concept of local grid refinement in critical areas. This technique is considerably more computationally complex but has great potential in future reservoir simulation efforts.

3. Development of representative model equations. Fluid motions in porous media are governed by the same fundamental laws that govern their flow in pipelines and rivers. These laws are based on conservation of mass, momentum, and energy. Additional governing equations which must be specified are rate equations, generally a form of Darcy's law, and equations of state. The governing equations model the basic processes which occur within the physical system. Since we usually do not have a complete knowledge of the total behavior of the system, a major difficulty in the modeling procedures is the choice of a set of governing equations which accurately describes the complex physical process.

The more complex the physical phenomenon, the more complicated the mathematical model must be, and thus the more difficult it will be to analyze and solve.

In addition to the governing equations, in order to have a complete mathematical model, we must specify a description of the reservoir, a set of initial conditions, and a set of appropriate boundary conditions to describe flow into or out of the reservoir. In this treatment, the boundary conditions will be separated into external boundary conditions and well specifications. For simulation of large reservoirs with several wells, the well models are a large source of error.

We first develop governing equations describing the flow rates of the fluid. Using these results we develop the equations for single-phase flow. With the definitions of relative permeabilities, we then generalize our mathematical models to multi-phase flow. Throughout the process of developing the governing equations for various flows, we consider mathematical properties of the equations which are derived, such as existence, uniqueness, and regularity of the solutions. Certain problems and future research areas will be pointed out throughout the development.

Perhaps the most widely used law or correlation which can be incorporated in analytical models of flow in porous media is Darcy's law, discovered in 1856 by the French engineer Henry D'Arcy. Although D'Arcy's experiments dealt only with laminar flow of water through various media, they established the basic relationship between the flow rate and the pressure gradient which can be modified in various ways to model a wide variety of flows in different regimes.

Darcy's law states that the volumetric flow rate Q of a homogeneous fluid through a porous medium is proportional to the pressure or hydraulic gradient and to the cross-sectional area A normal to the direction of flow and inversely proportional to the viscosity μ of the fluid. The law defines a concept of permeability k of the rock, which quantifies the ability of the rock to transmit fluid. We can write the superficial fluid velocity (Darcy velocity $\tilde{u}$) by

$$(3.1) \qquad \tilde{u} = \frac{Q}{A} = -\frac{k}{\mu}(\nabla p - \rho g \nabla Z)$$

where p is the fluid pressure, ρ is the fluid density, g is the magnitude of the acceleration due to gravity, the depth Z is a vector function of (x, y, z) pointing in the direction of gravity, and k is an absolute permeability tensor with units of darcies (length squared). In most uses of Darcy's law in reservoir simulation, it is assumed that k is the special diagonal tensor

$$(3.2) \qquad k = \begin{bmatrix} k_x & 0 & 0 \\ 0 & k_y & 0 \\ 0 & 0 & k_z \end{bmatrix},$$

where k_x, k_y, and k_z are interpreted as permeabilities in the x, y, and z directions, respectively. If $k_x = k_y = k_z$, the medium is termed isotropic; otherwise it is called anisotropic. In most flooding regimes, both the permeabilities and viscosities depend upon the saturations of the various phases. These dependencies form the basis of the relative permeability approximations to be discussed later in this section. Also in thermal recovery, all coefficients in Darcy's law require a temperature dependence.

Once the limitations and necessary modifications of Darcy's law in various flow regimes are understood, we can begin to derive the flow equations. The equation governing the single-phase flow of a fluid through a porous medium is developed by combining the following: 1) conservation of mass, 2) Darcy's law, and 3) an equation of state. If one considers a typical volume element, V, then conservation of mass implies that the rate of mass accumulated within V equals the rate of mass flow across the boundary of V plus the amount of mass injected into V via wells (sources or sinks). If ϕ, the porosity of the material, denotes the fraction of the volume V available for flow, ρ is the density per unit volume, ∂V is the boundary of V with normal vector ν, $\tilde{u}$ is the superficial Darcy velocity, and q is the mass flow rate per unit volume injected into (or produced from) V, then conservation of mass can be described by the equation

$$(3.3) \qquad \frac{d}{dt}\int_V \phi\rho(x,t)dx = -\int_{\partial V}\rho(x,t)\tilde{u}\cdot\nu ds + \int_V qdx.$$

If we use the divergence theorem to see that

$$(3.4) \qquad \int_{\partial V}\rho\tilde{u}\cdot\nu ds = \int_V \nabla\cdot(\rho\tilde{u})dx,$$

and interchange d/dt with spatial integration,

$$(3.5) \qquad \frac{d}{dt}\int_V \phi\rho(x,t)dx = \int_V \frac{\partial(\phi\rho)}{\partial t}dx,$$

we obtain

$$(3.6) \qquad \int_V \frac{\partial(\phi\rho)}{\partial t}dx = \int_V \left[-\nabla\cdot(\rho\tilde{u}) + q\right]dx.$$

Since equation (3.6) is to hold for any volume element V, we obtain the partial differential equation (pde) form

$$(3.7) \qquad \frac{\partial(\phi\rho)}{\partial t} = -\nabla\cdot(\rho\tilde{u}) + q$$

for single-phase flow subject to a superficial velocity $\tilde{u}$. Combining Darcy's law (3.1) with (3.7), we obtain

$$(3.8) \qquad \frac{\partial(\phi\rho)}{\partial t} = \nabla \cdot \frac{\rho k}{\mu}(\nabla p - \rho g \nabla Z) + q, \qquad x \in \Omega, \quad t \in [t_0, t_1],$$

where Ω is our spatial domain and $[t_0, t_1]$ is the time interval under consideration.

Note that (3.8) is a second-order pde in the two dependent variables p = pressure and ρ = density. By expressing p as a function of ρ, or vice versa, we can obtain a pde in only one independent variable. This is accomplished by use of an equation of state describing the relationship between p and ρ.

As an equation of state, we shall use the definition of fluid compressibility, c:

$$(3.9) \qquad c = -\frac{1}{V}\frac{\partial V}{\partial p}\bigg|_T.$$

at a fixed temperature T. Since density is equal to mass divided by volume, (3.9) can alternately be described by

$$(3.10) \qquad c = \frac{1}{\rho}\frac{\partial \rho}{\partial p}\bigg|_T.$$

Separating variables in (3.10) and denoting ρ_0 as the density at pressure p_0, we obtain

$$(3.11) \qquad p = p_0 + \frac{1}{c}\ln\left(\frac{\rho}{\rho_0}\right)$$

or, equivalently,

$$(3.12) \qquad \rho = \rho_0 \exp(c(p - p_0)).$$

Thus there is a one-to-one, invertible mapping between p and ρ described by (3.11) or (3.12). Therefore, using the chain rule, (3.8) can be written as a single pde for the density

$$(3.13) \qquad \frac{\partial(\phi\rho)}{\partial t} = \nabla \cdot \frac{k}{\mu c}(\nabla p - \rho^2 c g \nabla Z) + q, \qquad x \in \Omega, \quad t \in [t_0, t_1],$$

or as an equivalent, nonlinear pde for the unknown p.

Since (3.13) is a parabolic pde, it requires an initial specification of the density throughout the domain

$$(3.14) \qquad \rho(x, 0) = \rho_0(x), \qquad x \in \Omega,$$

and boundary conditions. Usually, Neumann boundary conditions, or the specification of the mass flow across the boundary, is used:

$$(3.15) \qquad \rho\tilde{u} \cdot \nu\big|_{\partial\Omega} = g_1(x, t), \qquad x \in \partial\Omega, \quad t \in [t_0, t_1].$$

However, in some situations when the density (or pressure) can be measured at the boundary, Dirichlet conditions are imposed.

If the compressibility of the fluid is small, as it is for many liquids, then the Taylor series expansion for the exponential in (3.12) can be truncated and the assumption can be made that

(3.16) $$\rho = \rho_0(1 + c(p - p_0)).$$

Under this assumption, a pde for pressure can be written from (3.13)

(3.17) $$c\frac{\partial(\phi\rho)}{\partial t} = \nabla \cdot \frac{k}{\mu}(\nabla p - \rho_0(1 + c(p - p_0))g\nabla Z)$$
$$+ c\nabla p \cdot (\nabla p - \rho_0(1 + c(p - p_0))g\nabla Z) + \frac{q}{\rho},$$
$$x \in \Omega, t \in [t_0, t_1].$$

Under the even stricter assumption that the fluid is *incompressible* ($c = 0$), we can obtain from (3.17) the elliptic pde

(3.18) $$0 = \nabla \cdot \frac{k}{\mu}(\nabla p - \rho_0 g \nabla Z) + \frac{q}{\rho}, \qquad x \in \Omega.$$

This is also the form of the equation for *steady-state* ($\partial(\phi\rho)/\partial t = 0$) flow. Thus for either incompressible fluids or steady-state flow, from (3.3) we see that if Neumann boundary conditions or mass flow rates are specified on all of the boundary, then for a solution to equation (3.18) to exist, we must have

$$\int_V q(x, t)dx = 0, \qquad t \in [t_0, t_1],$$

and flow into all of the injection wells must be exactly balanced by flow out of all of the production wells. Also for (3.18) with only flow boundary conditions, the pressure is only determined to within a constant; if $\tilde{p}$ satisfies (3.18), then $\tilde{p} + \tilde{c}$ satisfies (3.18) for any constant $\tilde{c}$. This lack of uniqueness causes no problems as long as it is understood and taken into account in the simulation.

At this point we note that through the derivation presented above, we passed from (3.6) holding for any arbitrary volume element V to (3.7) holding at any point in our domain. For (3.7) to hold mathematically, at an arbitrary point, q must be thought of as a "point" source or sink, or a Dirac delta function. This interpretation has been shown to be a good model for well flow behavior at some minimal distance away from the wells. However, this model does not describe flow at the well without some modification since it would yield unphysical "infinite" pressures and fluid velocities at the well (which is then assumed to have zero well-bore radius). For specification of mass flow rates at the wells, this model is generally adequate for simulation purposes since, for large reservoirs with many wells, the grid blocks are necessarily large enough that behavior "at the well-bore" is not required. However, if a bottom hole pressure is specified as a well-related boundary condition, the model must be modified to account for a "finite" and specified pressure.

When "point source" models are used in this fashion, the modeler must realize that Dirac delta sources and sinks affect the regularity of the solutions of the governing equations adversely in the neighborhood of the wells. Thus standard numerical approximations will not converge well (or at all for velocities) in the neighborhoods of the wells. This is an extremely important observation for those who use the numerical simulators in a history-matching process to obtain better coefficients for porosity and permeability for use in later simulation. For an overview of the history-matching process in reservoir simulation and how it is used to obtain better approximations of the global reservoir properties for use in the simulators, see a survey paper by the author [50] on determination of coefficients in reservoir model equations.

If point sources are used in the governing equations as well models, one can obtain the asymptotic behavior of p and u and ascertain roughly how fast they "blow up" as the point is approached. Functions with the same asymptotic singular behavior at the wells can then be subtracted from the solutions and the remaining, smoother part can be approximated numerically. Then the singular parts are added to these numerical results to approximate the true solutions. The computational success in subtracting out well singularities in reservoir problems has been reported in [32], [42], [43], [49], [54], [56], [57], [67], [96], [103], [104], [117]. In one special case of miscible displacement simulation [57] the analysis of this procedure has been coupled with reduced smoothness properties at the wells to obtain reduced-order convergence rates for the associated numerical approximations.

If we consider (3.6) as the governing equation instead of its counterpart (3.7), the "point source" interpretation is not so crucial as long as the well is not at the boundary of a computational element. Most of the simulation done in the petroleum industry today uses a different type of well model which can more easily accommodate the specification or computation of bottom hole pressures at the wells. These models are based on analytic solutions for steady-state radial flow equations and incorporate the size of the computational grid blocks and the true well-bore radius. The simplest versions of these models relate the pressure p at a distance r from the well to the volumetric flow rate q via the equation

$$(3.19) \qquad p_e - p = \frac{q\mu}{2\pi hk} \ln\left(\frac{r_e}{r}\right),$$

where p_e is the pressure at the "effective" drainage radius r_e, and h is the thickness of the region of assumed radial flow. Relationships between the effective radius r_e and the dimensions of the computational grid blocks and also the addition of other factors for well damage and other "skin effects" are discussed in detail in the literature [86], [89].

These well models are based on radial flow. In general, they do not account for the specific location of the well within the grid block and are thus a source of modeling error. In enhanced recovery processes where the flow at a production well is not radial, better well models are needed. Since most of the input and output of the recovery processes and their models relate directly to the wells, research in well models is crucial for improved simulation.

The single-phase flow equations developed above are not sufficient to model the simultaneous interactive flow of two or more phases; this type of flow dominates essentially all of the enhanced recovery processes. We next develop the basic equations for multi-phase flow in porous media. See also Peaceman [88].

We shall first assume that we have two fluid phases flowing simultaneously, that the fluids are immiscible, and that there is no mass transfer between the fluids. One fluid will wet the porous medium more than the other and will be termed the wetting phase fluid. The subscript convention will be w for the wetting phase and n for the nonwetting phase. The saturation of a phase is the fraction of the space available to flow occupied by that phase. Since both phases are flowing, we have

$$(3.20) \qquad S_w + S_n = 1.$$

We will also have separate pressures for each phase, p_w and p_n. The difference between these pressures is the capillary pressure, p_c. We shall assume that the capillary pressure is a unique nonlinear, monotone decreasing function of saturation:

$$(3.21) \qquad p_c(S_w) = p_n - p_w.$$

Each phase will have its own density, its own viscosity, and its own Darcy velocity. Analogues of (3.1) are

$$(3.22a) \qquad \tilde{u}_w = -\frac{k_w}{\mu_w}(\nabla p_w - \rho_w g \nabla Z),$$

$$(3.22b) \qquad \tilde{u}_n = -\frac{k_n}{\mu_n}(\nabla p_n - \rho_n g \nabla Z),$$

where $\tilde{u}_w$ and $\tilde{u}_n$ are the superficial velocities, k_w and k_n are the "effective" permeabilities. μ_w and μ_n are the viscosities, p_w and p_n are the pressures, and ρ_w and ρ_n are the densities for the wetting and nonwetting phases, respectively. In the simultaneous flow of two fluids, each fluid interferes with the flow of the other and the effective permeabilities are less than or equal to the single-phase permeability k of the porous medium. We then define *relative permeabilities* as

$$\text{(3.23a)} \qquad k_{rw} = \frac{k_w}{k} \leq 1,$$

$$\text{(3.23b)} \qquad k_{rn} = \frac{k_n}{k} \leq 1,$$

from zero to one. We note that if the two fluids are miscible and flow as one phase then there is no interference of flow between the phases and $k_{rw} = k_{rn} = k$. In this case, the relative permeabilities are independent of saturation in any region of miscibility.

Since, in the immiscible case, we have assumed there is no mass transfer between phases, we will have governing equations describing conservation of mass within each phase. The same arguments which led to (3.8) will then yield the system, for $x \in \Omega$ and $t \in [t_0, t_1]$,

$$\text{(3.24a)} \qquad \frac{\partial(\phi \rho_w S_w)}{\partial t} = \nabla \cdot \frac{\rho_w k k_{rw}}{\mu_w} (\nabla p_w - \rho_w g \nabla Z) + q_w,$$

$$\text{(3.24b)} \qquad \frac{\partial(\phi \rho_n S_n)}{\partial t} = \nabla \cdot \frac{\rho_n k k_{rn}}{\mu_n} (\nabla p_n - \rho_n g \nabla Z) + q_n,$$

where q_w and q_n are the mass flow rates of the wetting and nonwetting phases, respectively. The equations are coupled via the constraints (3.20) and (3.21) and the pressures and densities for each phase are related by equations of state of the form

$$\text{(3.25a)} \qquad c_w = \frac{1}{\rho_w} \frac{\partial \rho_w}{\partial p_w}\bigg|_T,$$

$$\text{(3.25b)} \qquad c_n = \frac{1}{\rho_n} \frac{\partial \rho_n}{\partial p_n}\bigg|_T.$$

Since the system (3.24) resembles equation (3.8) closely, one might think that (3.24) is a system of parabolic pde's with diffusion-like properties. This is not necessarily the case. The character of the system (3.24) can be seen after using a transformation suggested by Chavent [13], [14]. One can see the properties more clearly under the assumptions that the densities of the phases and the porosity are constants and there are no gravity terms. Then defining the total and phase mobilities as

$$\text{(3.26a)} \qquad \lambda(S_w) = \frac{k_{rn}(S_w)}{\mu_n} + \frac{k_{rw}(S_w)}{\mu_w},$$

$$\text{(3.26b)} \qquad \lambda_i(S_w) = \frac{k_{ri}(S_w)}{\lambda(S_w)\mu_i}, \qquad i = n, w,$$

defining the average pressure [14] as

$$p = \frac{1}{2}(p_w + p_n) + \frac{1}{2}\int_0^{p_c}(\lambda_n(\xi) - \lambda_w(\xi))d\xi, \tag{3.27}$$

and defining the total fluid velocity as

$$\tilde{u} = -k(x)\lambda(S_w)\nabla p, \tag{3.28}$$

we can add and subtract the equations in (3.24) and collect terms as in [39] to obtain, for $x \in \Omega, t \in [t_0, t_1]$,

$$\nabla \cdot \tilde{u} \equiv -\nabla \cdot (k(x)\lambda(S_w)\nabla p) = \frac{q_n}{\rho_n} + \frac{q_w}{\rho_w}, \tag{3.29a}$$

$$\phi\frac{\partial S_w}{\partial t} - \nabla \cdot (k\lambda\lambda_n\lambda_w\frac{\partial p_c}{\partial S_w}\nabla S_w) + \lambda_w\tilde{u}\cdot\nabla S_w = f(q_w, p_w, \lambda_w, q_n, p_n, \lambda_n), \tag{3.29b}$$

where f is a simple linear function of the flow properties.

The pressure equation (3.29a) is an elliptic pde. The assumptions made to obtain (3.29a), that the densities of the phases and the porosity were constant, is akin to an assumption of incompressibility of both the fluids and the rock. Without this assumption an equation of the form

$$\nabla \cdot k\lambda\nabla p = \left(\frac{q_n}{\rho_n} + \frac{q_w}{\rho_w}\right) + \phi c_t \frac{\partial p}{\partial t} \tag{3.30}$$

can be obtained [88] where c_t is the total compressibility of the system,

$$c_t = \frac{1}{\phi}\frac{\partial \phi}{\partial p} + (S_n c_n + S_w c_w). \tag{3.31}$$

If either fluid or the rock is compressible, then (3.29a) becomes (3.30) which is a parabolic pde. In many cases c_t is extremely small and (3.30) is either elliptic or almost elliptic.

The equation for the saturation, (3.29b), is similar to a nonlinear convection-diffusion equation. In many cases, capillary pressure effects are small and the second term on the left side of (3.29b) is small. Also, if $S_i = 0$, then $\lambda_i = 0$ for $i = w, n$, so the coefficients of the diffusion term degenerate if $S_w = 0$ or $S_n = 0$ (this happens near the injection well and in parts of the reservoir where the injected fluid has not reached). In these cases, the convection term, the third term on the left side of (3.29b), parabolic equations have many of the same properties as first-order hyperbolic equations, such as travelling wave fronts. The interface between the two fluids will move through the medium in a traveling front mode for a time. The front will be slightly diffused due to capillary pressure effects. Later, due to viscosity differences, fingering could destroy the frontal profile.

We have now developed governing equations for immiscible two-phase flow in a porous medium. If the fluids mix together and form one single phase which flows as one fluid, the displacement is termed fully miscible. The governing equations for the miscible displacement of one incompressible fluid by another in a porous medium is given by [35], [42], [55], [88]

$$\nabla \cdot \tilde{u} \equiv -\nabla \cdot \frac{k(x)}{\mu(S)} \nabla p = q, \tag{3.32a}$$

$$\phi \frac{\partial S}{\partial t} - \nabla \cdot (D \nabla S) + \tilde{u} \cdot \nabla S = (\tilde{S} - S)q, \tag{3.32b}$$

for $x \in \Omega, t \in [t_0, t_1]$, where p and $\tilde{u}$ are the total fluid pressure and velocity, S is the saturation of the invading fluid, $\tilde{S}$ is the specified injection or resident production saturation of the injected fluid, and q is the total volumetric flow rate at the well. D is the diffusion-dispersion tensor given by [55], [90]

$$(D_{ij}(x, \tilde{u}) = \phi d_m I + \frac{d_l}{|\tilde{u}|} \begin{pmatrix} u_1^2 & u_1 u_2 \\ u_1 u_2 & u_2^2 \end{pmatrix} + \frac{d_t}{|\tilde{u}|} \begin{pmatrix} u_2^2 & -u_1 u_2 \\ -u_1 u_2 & u_1^2 \end{pmatrix}, \tag{3.33}$$

where $\tilde{u} = (u_1, u_2)$, $|\tilde{u}|$ is the Euclidean norm of $\tilde{u}$, d_m is the molecular diffusion coefficient, and d_l and d_t are the magnitudes of longitudinal and transverse dispersion. Equations (3.32) will be derived in Chapter II of this volume.

As before, if the fluids are compressible, the elliptic pde (3.32) will become parabolic. In general, d_m is assumed to be very small with d_l and d_t somewhat larger. Since the magnitudes of the last two terms in D are approximately $d_l|\tilde{u}|$ or $d_t|\tilde{u}|$, we see more dispersive mixing where the velocities are higher, around the wells, and less out in the reservoir. Therefore, although the systems (3.29) and (3.32) appear to be very similar, there are some important differences. Both are transport dominated but not purely hyperbolic systems.

The equations used for simulators involving compositional effects, or mass transfer between phases without always being fully miscible, are related to those of immiscible and miscible displacement, being capable of modeling aspects of both regimes. Since mass is exchanged between phases, mass is not conserved within each phase as in immiscible flow. Instead, there are several components flowing simultaneously, perhaps in several phases, and we assume that the total mass of each component is conserved. For example, if the phases are denoted o, g, and w for oil, gas and water, and if C_{io}, C_{ig}, and C_{iw} denote the mass fraction of the ith component in the respective phases, then the mass flux density for the ith component can be denoted by

$$C_{ig} \rho_g \tilde{u}_g + C_{io} \rho_o \tilde{u}_o + C_{iw} \rho_w \tilde{u}_w$$

where ρ_i and $\tilde{u}$, $i = g, o, w$, are the phase densities and Darcy velocities. Then the conservation of mass in the ith component can be described by

$$\frac{\partial}{\partial t}[\phi(C_{ig}\rho_g S_g + C_{io}\rho_o S_o + C_{iw}\rho_w S_w)]$$
(3.34)
$$= q_i + \nabla \cdot [C_{ig}\rho_g \tilde{u}_g + C_{io}\rho_o \tilde{u}_o + C_{iw}\rho_w \tilde{u}_w], \qquad i = 1, \ldots, N,$$

where N is the total number of components treated. Coupled to these conservation equations are a complex phase package which, given pressure-volume-temperature data, describes relative phase equilibria, Darcy's law for the flow of each phase, mole fraction and saturation constraints, and either a Peng–Robinson [91] or Redlich–Kwong cubic equation of state. For examples of compositional simulators, see [23], [24], [51], [71], [72], [83], [84], [120]. The component continuity equations (3.34) have mathematical properties similar to those of both (3.29) and (3.32), with transport dominance properties.

Chemical simulation models are, mathematically, very similar to compositional models. Although oil is usually a single component, other components such as surfactant, polymer, calcium ion, total anions, and water must be accounted for. The surfactant and polymer slugs are usually quite small and govern most of the physics, so physical dispersion monitoring and numerical dispersion control is essential. Usually four phases are modeled: an immobile rock phase and three mobile phases. The mobile phases are termed aqueous (predominantly water), oleic (predominantly oil) and microemulsion (containing most of the surfactant). Ionic transfer from the rock phase is crucial to the interfacial activity. The conservation equations, essentially of the form (3.34), are coupled to complex ionic transfer relations and phase equilibria equations. For example equations from a prototype chemical simulator, see [25], [29].

For thermal simulation, along with balancing saturations of oil, water, oxygen, inert gasses, and total fluid, a total energy balance is used. This equation, which relates heat conduction, transport of enthalpy, heat of combustion, and energy input at wells to the increase in internal energy of both the fluids and the rock formation, again is a nonlinear transport dominated diffusion equation. The combustion equation relates the stochiometric coefficients of oil, oxygen, water, and inert gasses which, in turn, contribute to sink terms in the mass balance equations for these elements. For more details and example equations for thermal simulation, see [19], [21], [30], [68], [97], [114].

Although the details of each of the simulators described above involve different complexities, the basic flow equations are very similar, mathematically. Some of the common mathematical, numerical approximation, and computational difficulties such as numerical dispersion, grid orientation problems, inaccurate darcy velocities, stability, computational complexity, and modeling of viscous fingering are described in the next section. Several of these topics are then treated in more detail in later chapters of this volume.

4. Numerical difficulties in simulation. In this section we shall describe some of the numerical difficulties which have plagued numerical reservoir simulation over the last few years. Some of these difficulties are well enough understood that they are now beginning to be overcome. Others are important areas for future research.

As we noted in §3, many of the partial differential equations being used as models for reservoir simulation are transport-dominated convection-diffusion equations in several space dimensions. Finite difference spatial discretization techniques are used predominantly in the petroleum industry for reservoir simulation.

Let $\{x_0, x_1, \cdots, x_N\}$ be a uniform subdivision of a spatial interval $[x_0, x_N]$ with mesh length Δx and denote $S_i = S(x_i)$ for $i = 0, 1, \cdots, N$. It is well known that centered differencing of any first derivative transport term, given by

$$(4.1) \quad \frac{S_{i+1} - S_{i-1}}{2\Delta x} = \frac{\partial S}{\partial x}(x_i) + \frac{(\Delta x)^2}{6} \frac{\partial^3 S}{\partial x^3}(\eta), \quad x_{i-1} \leq \eta \leq x_{i+1},$$

has $O((\Delta x)^2)$ accuracy if S is sufficiently smooth, but causes some stability problems. A common stabilizing ploy is to use a one-sided difference approximation in the direction of fluid flow instead of the central difference quotient. From simple Taylor series truncation analysis, one can show that

$$(4.2) \quad \frac{S_i - S_{i-1}}{\Delta x} = \frac{\partial S}{\partial x}(x_i) - \frac{\Delta x}{2} \frac{\partial^2 S}{\partial x^2}(\eta), \quad x_{i-1} \leq \eta \leq x_i,$$

and this method has only $O(\Delta x)$ accuracy. In addition, similar analysis shows that

$$(4.3) \quad \begin{aligned} \frac{S_i - S_{i-1}}{\Delta x} &= \frac{S_{i+1} - S_{i-1}}{2\Delta x} - \frac{\Delta x}{2} \left[\frac{S_{i+1} - 2S_i + S_{i-1}}{(\Delta x)^2} \right] \\ &= \frac{\partial S}{\partial x}(x_i) - \frac{\Delta x}{2} \frac{\partial^2 S}{\partial x^2}(x_i) + O((\Delta x)^2). \end{aligned}$$

Thus one-sided differencing of the transport term, commonly called upwind differencing or upstream weighting, can be thought of as being an $O((\Delta x)^2)$ approximation to

$$\frac{\partial S}{\partial x}(x_i) - \frac{\Delta x}{2} \frac{\partial^2 S}{\partial x^2}(x_i)$$

instead of $(\partial S/\partial x)(x_i)$. Since $\partial^2 S/\partial x^2$ is a physical diffusion-like term, this in essence amounts to the introduction of an "artificial diffusion" of size $\Delta x/2$.

This artificial diffusion or numerical dispersion does stabilize the difference method (by moving an off-diagonal term in the matrix used to solve the problem onto the diagonal to give more diagonal dominance and a guaranteed solution) but the numerical error induced can be disastrous to reservoir simulation. In particular this diffusion, the magnitude of which depends upon the mesh spacing Δx, will often artificially smear out sharp fronts or fluid interfaces and destroy the physical information in the model. See Fig. 2. A dramatic example of the disastrous effects of this diffusion can be seen in thermal simulation of a burning front which travels through the reservoir. The width of the front may be very small compared to the grid size which is economically possible for simulation. The temperature distribution will then be a narrow pulse with very large values. Numerical dispersion will smear the pulse out, increasing the width and drastically decreasing the largest predicted temperature values. Below certain critical temperatures, the physical combustion will cease and no more thermal energy will be generated. If the computed temperature distribution is artificially diffused to obtain temperatures below this critical combustion temperature, the model will predict a cease in the burning when, in fact, combustion is still occurring. Thus the physics of the process will be lost and the simulation will be worthless. Similar disastrous loss of modeling of the physics occurs if a surfactant or polymer pulse is artificially diffused in the simulation process. Since the physics at the fluid interface governs much of enhanced recovery simulation, these interfaces cannot be artificially dispersed without destroying the simulation process.

From (4.3) we see that the numerical dispersion induced by upstream weighting depends upon the mesh spacing. If a very small Δx is used, this dispersion can be controlled. Local mesh refinement around fronts in one space dimension is a viable control of numerical dispersion. In two or three space dimensions local grid refinement which adaptively follows a moving interface is

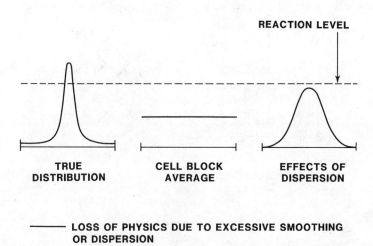

FIG. 2. *Front resolution—thermal simulation.*

an important topic of future research. A first step in this direction is presented in [33].

Another disastrous effect of upstream weighting of the transport term is a phenomenon of obtaining drastically different numerical predictions from simulators due only to different spatial orientations of the computational grid. If upstream weighting is used for transport terms in each of the x, y, and z directions, one will obtain an artificial dispersion term of the form

$$(4.4) \quad d = -\frac{\Delta x}{2}\frac{\partial^2 S}{\partial x^2} - \frac{\Delta y}{2}\frac{\partial^2 S}{\partial y^2} - \frac{\Delta z}{2}\frac{\partial^2 S}{\partial z^2}$$

which is not rotationally invariant and is thus directionally dependent. When modeling multiphase processes with high mobility ratios (basically large viscosity ratios), once a preferential flow path has been established, the greater mobility of the less viscous fluid causes this flow path to dominate the flow pattern. With standard finite difference methods utilizing five-point (in two space dimensions) or seven-point (in three space dimensions) difference stars, slightly preferred flow paths are established along the coordinate directions. Then the use of a stabilizing technique, such as upstream weighting of the flow terms, which is not rotationally invariant, greatly enhances flow in these preferred directions. This grid orientation effect is dramatic in cases with very high mobility ratios. See Fig. 3 for an illustration of how grids can often have different orientations, even with uniform five-spot well configurations. Figure 4 then illustrates the very different flow patterns with parallel, or diagonal grid orientations. The parallel grid will predict earlier breakthrough of the injected fluid at the production wells while the diagonal grid will predict much greater sweep efficiency. In Fig. 5 one can see that these two orientations predict very different total recovery curves with time

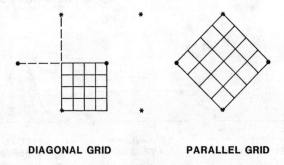

DIAGONAL GRID   **PARALLEL GRID**

* = INJECTION WELLS
• = PRODUCTION WELLS

FIG. 3. *Grid orientation effect—uniform five-spot pattern.*

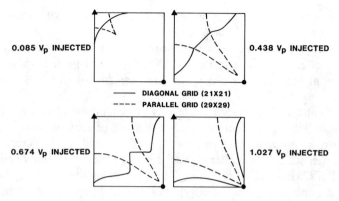

FIG. 4. *Concentration contours in a five-spot, M = 10. (From Yanosik and McCracken [118].)*

and thus very different economics of the enhanced recovery process. Although the parallel orientation is closer to the true flow pattern, due to the standard uneven placement of many wells in a large simulation, one cannot orient the grid in a preferred way for all wells. Therefore, for reliable simulation this difficulty must be overcome.

Although nine-point finite difference stars in two space dimensions give a better initial flow distribution [118], standard upstream weighting still causes severe grid orientation effects since it is not a rotationally invariant operator. Other authors have used operators which are more nearly rotationally invariant to stabilize the computations without introducing severe grid-orientation effects. Potempa [96] has developed a very robust scheme which has been used to stabilize difficult steam-flood calculations without introducing significant grid-

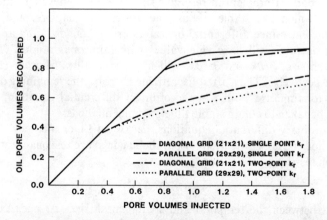

FIG. 5. *Predicted recovery performance, M = 10. (From Yanosik and McCracken [118].)*

orientation effects. However, this technique is still somewhat dispersive and produces some smearing of sharp temperature distributions. A delicate balance between stability and excessive diffusiveness must be maintained while grid orientation effects are being controlled. Potempa's scheme and similar rotationally invariant techniques are discussed in more detail in Chapter II of this volume.

In most enhanced recovery processes, a physical dispersion is present which helps to stabilize the flow until it reaches the level of unstable viscous fingering. This dispersion is due to mechanical mixing of the fluids, at a microscopic level, or, on a larger scale, to macroscopic variations in rock permeability and porosity. Section 2 of Chapter II gives an excellent description of the origin, magnitude, importance, and predominance of types of physical dispersion together with an extensive review of the engineering and mathematical literature on the subject. Arguments are made for why this phenomenon should be incorporated in the mathematical models for enhanced recovery. As noted in Chapter II, this physical dispersion is often sufficient to stabilize the mathematical computations by itself. It is also modeled via a dispersion tensor which is rotationally invariant, hence causing no grid orientation effects. When this physical dispersion is included in the mathematical model, often no artificial dispersion need be introduced by upstream weighting or other techniques, and sharp fluid interfaces can be computed with essentially no grid orientation effects. Discussion of various finite element techniques [15], [32], [38], [47], [53], [54], [102], [104], [105] which have utilized only physical dispersion levels to simulate difficult enhanced recovery methods appear in Chapter II of this volume. See Fig. 6 for typical recovery curves of these methods for high mobility ratios.

Since the models for essentially all the enhanced recovery processes involve large, coupled systems of nonlinear, time-dependent partial differential equations, an important problem in reservoir simulation is to develop time-stepping procedures which are stable, robust, accurate, self-adaptive, and computationally efficient. Since differential operators are unbounded, any approximate numerical operator will have eigenvalues with large magnitudes. (If Δx is a spatial mesh size, $\partial/\partial x$ and $\partial^2/\partial x^2$ will have eigenvalues of order $(\Delta x)^{-1}$ and $(\Delta x)^{-2}$, respectively). Thus, after discretizing in space, the remaining differential equations for time are "stiff" in the ordinary differential equation sense, and care must be taken to obtain stable time-stepping methods.

As for ordinary differential equations, backward Euler methods or modified Euler methods (Crank–Nicolson for parabolic pde's) are reasonably stable while a time-step restriction of the form

(4.5) $$\Delta t \leq K(\Delta x)^2$$

must hold between the temporal and spatial mesh sizes Δt and Δx for some constant K for stability to be assured with a forward Euler method. On the other

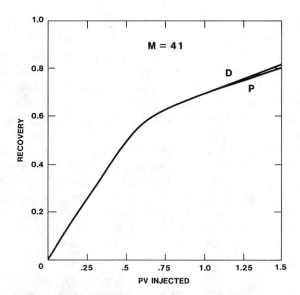

FIG. 6. *Recovery curves, high mobility ratios, M* = 41.

hand, a forward Euler or "explicit" method is a local computational scheme which is very efficient and a backward Euler or "implicit" method requires the solution of very large systems of nonlinear equations at each time step. Due to the computational complexity of linearizing via some Newton-like method and then solving, by direct elimination or iterative linear solution techniques, an extremely large set of equations at each time step, explicit methods have been used frequently in reservoir simulation. The stability condition (4.5), however, often requires excessively small time steps and enormous computations to simulate long time periods (several years) in a field-scale model. Problems with strong nonlinearities eventually convinced the industry that fully explicit methods could not be used efficiently for many problems.

A variation to obtain better stability without increasing computational complexity too much, termed IMPES [5], [88], [99], was then developed. Using (3.32) as a sample system, IMPES solves implicitly for pressure from (3.32a) and then explicitly for saturation from (3.32b), hence the name. This technique is still used widely in the industry and works well for problems of intermediate to easy difficulty. However, for difficult nonlinearities, the time step restriction (4.5) often imposes extremely small time steps and is not efficient enough.

The other extreme from a fully explicit time-stepping method is a fully coupled, fully implicit method which solves all of the coupled nonlinear pde's simultaneously in an implicit fashion. This is a very stable method and allows quite large time steps while maintaining stability. However, often the robustness of this method encourages the user to take time steps which are much too large for accuracy considerations and an artificial diffusion from the temporal truncation is introduced. Also, for the more complex simulators, where many equations must be coupled (e.g. thermal and compositional models), if a field-scale problem is addressed, the size of the matrices to be inverted quickly becomes too large, even for the super computers. These techniques are in use for several simulators today [6], [21], [85], [112] and research is underway to extend their capabilities to the more complex processes.

A variety of methods for sequentially solving the governing equations in an implicit manner without the full coupling have also been developed. These methods have less stability but better computational features than fully coupled, fully implicit methods and more stability and somewhat worse computational properties than IMPES. In many cases, the increased stability, which allows larger time steps than IMPES, actually allows less work because far fewer time steps and thus fewer computations are involved. Again, accuracy constraints should then be applied to insure that the accuracy is not destroyed with the larger time steps. The only rigorous asymptotic analysis for the coupled miscible displacement system assumed a sequential time-stepping procedure [51], [110]. The controversy over the relative merits of the various time-stepping methods continues to grow. The majority of the research effort in the last few years in the reservoir simulation field has concerned the stability of time-stepping methods, and the efficient linearization and iterative solution of the resulting equations. This author feels that robustness and efficiency have been considered so important that accuracy has been largely ignored. He feels that unless the discretizations are sufficiently accurate, it makes no sense to solve them robustly and efficiently to obtain meaningless results.

One major accuracy problem in reservoir simulation is the way that the Darcy velocities, which govern the basic flow properties of the fluids, are calculated. The Darcy velocity is a fairly smooth physical quantity. Recall that its definition, ignoring gravity terms, is

$$(4.6) \qquad \tilde{u} = -\frac{k}{\mu}\nabla p$$

from (3.1). Often the flow properties of the porous media change abruptly with sharp changes in lithology. Also across fluid interfaces the viscosity changes quite rapidly in space. These sharp changes are accompanied by large changes in the pressure gradient which, in a compensatory fashion, yield a fairly smooth Darcy velocity of the fluid. Standard finite difference or finite element procedures for solving a coupled system for some model determine the pressure p as a set of cell averages, nodal values, or piecewise smooth functions. The resulting p is then differenced or differentiated and then multiplied by a possibly rough function k/μ to determine the velocity $\tilde{u}$. These processes produce a rough and often inaccurate approximation of $\tilde{u}$ which then reduces the accuracy of the approximation of the various fluid saturations through the mass balance equations.

Recall that from physical considerations, (3.32 a) was originally the coupled system consisting of (4.6) and

(4.7) $$\nabla \cdot \tilde{u} = q.$$

These equations were coupled to form the second-order elliptic equation (3.32 a) for the pressure. Therefore, it seems only natural to solve the coupled system (4.6) and (4.7) simultaneously for the two physical quantities p and $\tilde{u}$. This is the motivation for mixed finite element methods. In solving for $-(k/\mu)\nabla p$ as one variable, we minimize the difficulties caused by rough coefficients k/μ in standard methods. In this way, the Darcy velocities can be obtained in a provably more accurate fashion. For analyses of these methods see [9], [10], [42], [43], [58], [70], [100], [117]. Versions of these methods have been successfully applied with other finite element methods to miscible displacement problems in [15], [42], [43], [70] and are being combined with finite difference methods in compositional simulation [51]. See Chapter II for an extensive discussion of the various recent results obtained using mixed finite element methods for miscible displacement problems.

An additional difficulty in the numerical modeling of enhanced recovery processes is the need to incorporate in our simulation models the bypassing of oil due to a viscous fingering phenomenon. We emphasize that the governing equations derived in §3 were obtained via a volume averaging mechanism which does not model physical behavior on a pore-volume scale which causes the viscous fingering effects. Since the mathematical equations developed in §3 are not capable of describing the physics of the instabilities on the microscopic level, they should not be expected to model fingering on that level. If, however, anisotropic rock properties are described on a large enough scale that they can be incorporated in the permeabilities on a grid-size level, then the results of the mathematical model should reflect these anisotropies in the form of a macroscopic fingering phenomenon due to varying flow velocities. If the mathematical model further includes differences in longitudinal versus transverse dispersion levels from (3.33), then fingers initiated by variable permeabilities should propagate and grow in a manner akin to viscous fingering on a smaller scale. If no permeability variations or longitudinal dispersion effects are included in the mathematical model, any fingering phenomenon noticed would be due to numerical errors and not the modeling of any physics. An example of fingering induced by a cell-sized permeability variation and enhanced by longitudinal dispersion effects which were incorporated in the model appears in Fig. 7. These computations were performed by Russell, Wheeler, and the author and are described in [54].

Analysis must be done on equations which describe the microscopic physics of

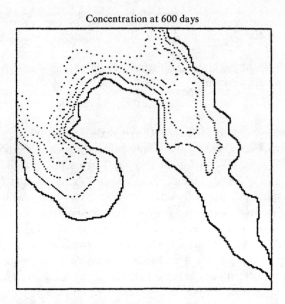

Concentration at 600 days

FIG. 7. *Random permeabilities—modified method of characteristics and mixed methods*, 20×20 *grid*, $M = 100$.

multiphase flow to understand the conditions under which the small flow perturbations caused by tortuous flow in the porous media will become unstable and will grow into large fingers which affect flow on the macroscopic or grid-size level. The understanding of this instability is crucial, both in attempts to stabilize the flow via polymers, etc., and to model the fingering phenomenon when it does occur. Knowledge of the conditions causing the onset of the instability is one major goal. Then the understanding of the nonlinear effects which cause certain fingers to grow preferentially and coalesce to form larger fingers is essential. The rate of this finger growth must be determined. Once the growth of the fingers is understood, from a microscopic to a macroscopic level, statistical methods can be used to incorporate the large-scale effects of fingering into the mathematical models used to simulate the large process. Methods for modeling the fingering phenomenon are presented in Chapter IV of this volume and in related papers [18], [61], [63], [64].

Acknowledgments. The author would like to thank Jim Douglas, Jr., Mary Wheeler, Tom Russell, Richard Kendall, Bob Heinemann, Alvis McDonald, Aziz Odeh, Larry Young, Don Peaceman, and John Wheeler for many valuable discussions concerning this work. He is indebted to Mobil Research and Development Corporation and to the University of Wyoming, where he was supported as the J. E. Warren Distinguished Professor of Energy and Environment.

REFERENCES

[1] N. ALBRIGHT, C. ANDERSON, AND P. CONCUS, *The random choice method for calculating fluid displacement in a porous medium*, in Boundary and Interior Layers—Computational and Asymptotic Methods, J. J. H. Miller, ed., Boole Press, Dublin, 1980, pp. 3–13.

[2] N. ALBRIGHT AND P. CONCUS, *On calculating flows with sharp fronts in a porous medium*, in Fluid Mechanics in Energy Conservation, J. Buckmaster, ed., Society for Industrial and Applied Mathematics, Philadelphia, 1980, pp. 172–184.

[3] N. ALBRIGHT, P. CONCUS, AND W. PROSKUROWSKI, *Numerical solution of the multidimensional Buckley–Leverett equation by a sampling method*, SPE 7681, 5th SPE Symposium on Reservoir Simulation, Denver, 1979.

[4] C. ANDERSON AND P. CONCUS, *A stochastic method for modeling fluid displacement in petroleum reservoirs*, in Analysis and Optimization of Systems, A. Bensoussan and J. L. Lions, eds., Lecture Notes in Control and Information Science 28, Springer-Verlag, Berlin, 1980, pp. 827–841.

[5] K. AZIZ AND A. SETTARI, *Petroleum Reservoir Simulation*, Applied Science Publishers, London, 1979.

[6] P. L. BANSAL, J. L. HARPER, A. E. MCDONALD, E. E. MORELAND, A. S. ODEH, AND R. H. TRIMBLE, *A strongly coupled fully implicit, three-dimensional, three-phase reservoir simulator*, SPE 8329, 54th SPE Annual Technical Conference, Las Vegas, 1979.

[7] J. BEAR, *On the tensor form of dispersion in porous media*, J. Geophys. Res., 66 (1961), pp. 1185–1197.

[8] ———, *Dynamics of Fluids in Porous Media*, American Elsevier, New York, 1972.

[9] F. BREZZI, *On the existence, uniqueness, and approximation of saddle-point problems arising from Lagrangian multipliers*, RAIRO Anal. Numer., 2 (1974), pp. 129–151.

[10] D. C. BROWN, *Alternating-direction iterative schemes for mixed finite element methods for second order elliptic problems*, Ph.D. thesis, Univ. Chicago, Chicago, 1982.

[11] S. E. BUCKLEY AND M. C. LEVERETT, *Mechanism of fluid displacements in sands*, Trans. AIME, 146 (1942), pp. 107–116.

[12] V. CASULLI AND D. GREENSPAN, *Numerical simulation of miscible and immiscible fluid flow in porous media*, Soc. Pet. Eng. J., 22 (1982), pp. 635–646.

[13] G. CHAVENT, *A new formulation of diphasic incompressible flows in porous media*, Lecture Notes in Mathematics 503, Springer-Verlag, Berlin, 1976, pp. 258–270.

[14] ———, *The global pressure, a new concept for mobilization of compressible two phase flow in porous media*, in Flow and Transport in Porous Media, A. Verruyt and F. B. J. Barends, eds., Balkema, Rotterdam, 1981.

[15] G. CHAVENT, J. JAFFRE, G. COHEN, M. DUPUY AND I. DIESTE, *Simulation of two-dimensional waterflooding using mixed finite element methods*, SPE 10502, 6th SPE Symposium on Reservoir Simulation, New Orleans, 1982, pp. 147–158.

[16] R. L. CHUOKE, P. VAN MEURS, AND C. VAN DER POEL, *The instability of slow, immiscible, viscous liquid liquid displacements in permeable media*, Trans. AIME, 216 (1959), pp. 188–194.

[17] A. J. CHORIN, *Random choice solution of hyperbolic systems*, J. Comp. Phys., 22 (1976), pp. 517–533.

[18] ———, *Instability of fronts in a porous medium*, Report LBL-15893, Lawrence Berkeley Lab., Univ. California, Berkeley, 1983.

[19] C. CHU, K. H. COATS, W. D. GEORGE, AND B. E. MARCUM, *Three-dimensional simulation of steam flooding*, SPE J., Trans. AIME, 14 (1974), pp. 573–592.

[20] E. L. CLARIDGE, *Prediction of recovery in unstable miscible flooding*, Soc. Pet. Eng. J., 12 (1972), pp. 143–155.

[21] K. H. COATS, *A fully implicit steamflood model*, Soc. Pet. Eng. J., 18 (1978), pp. 369–383.

[22] ———, *Reservoir simulation: state of the art*, J. Pet. Tech., 34 (1982), pp. 1633–1642.

[23] K. H. COATS AND A. B. RAMESH, *Effect of grid type and difference scheme on pattern steamflood simulation results,* SPE 11079, 6th SPE Symposium on Reservoir Simulation, New Orleans, 1982.
[24] K. H. COATS, *An equation of state compositional model,* Soc. Pet. Eng. J., 20 (1980), pp. 363–375.
[25] M. F. COHEN, *Numerical dispersion in chemical-waterflooding simulation—a comparison among finite element and finite difference methods,* Comp. Methods Appl. Mech Eng., to appear.
[26] P. COLELLA, *Glimm's method for gas dynamics,* SIAM J. Sci. Statist. Comput., 3 (1982), pp. 76–110.
[27] R. E. COLLINS, *Flow of Fluids through Porous Materials,* Van Nostrand-Reinhold, New York, 1961.
[28] P. CONCUS, *Calculation of shocks in oil reservoir modeling and porous flow,* in Numerical Methods for Fluid Dynamics, K. W. Morton and M. J. Baines, eds., Academic Press, New York, 1982, pp. 165–178.
[29] P. CONCUS AND W. PROSKUROWSKI, *Numerical solution of a nonlinear hyperbolic equation by the random choice method,* J. Comp. Phys., 30 (1979), pp. 153–166.
[30] R. B. CROOKSTON, W. E. CULHAM AND W. H. CHEN, *Numerical simulation model for thermal recovery processes,* Soc. Pet. Eng. J., 19 (1979), pp. 37–58.
[31] B. L. DARLOW, *A penalty-Galerkin method for solving the miscible displacement problem,* Ph.D. thesis, Rice Univ., Houston, 1980.
[32] B. L. DARLOW, R. E. EWING, AND M. F. WHEELER, *Mixed finite element methods for miscible displacement in porous media,* SPE 10501, 6th SPE Symposium on Reservoir Simulation, New Orleans, 1982, pp. 137–145; Soc. Pet. Eng. J., to appear.
[33] J. C. DIAZ, R. E. EWING, R. W. JONES, A. E. MCDONALD, L. M. UHLER AND D. U. VON ROSENBERG, *Self-adaptive local grid refinement in enhanced oil recovery,* Proc. 5th International Symposium on Fluid Flow Problems, Austin, TX, 1983.
[34] E. L. DOUGHERTY, *Mathematical model of an unstable miscible displacement,* Soc. Pet. Eng. J., 3 (1963), pp. 155–163.
[35] J. DOUGLAS, JR., *The numerical solution of miscible displacement in porous media,* in Computational Methods in Nonlinear Mechanics, J. T. Oden, ed., North-Holland, Amsterdam, 1980, pp. 225–238.
[36] ———, *Simulation of a linear waterflood,* in Free Boundary Problems, Vol. II. Istituto Nazionale di Alta Matematica "Trancesco Severi," Roma, 1980.
[37] ———, *Recent results concerning simulation of miscible flow in porous media,* Seminar on Numerical Analysis and Its Applications to Continuum Physics, Calecao, Atas, Vol. 12, Rio de Janeiro, 1980.
[38] ———, *Simulation of miscible displacement in porous media by a modified method of characteristics procedure,* in Numerical Analysis, Dundee, 1981, Lecture Notes in Mathematics 912, Springer-Verlag, Berlin, 1982.
[39] ———, *Finite difference methods for two-phase, incompressible flow in porous media,* SIAM J. Numer. Anal., 20 (1983), pp. 681–696.
[40] J. DOUGLAS, JR., B. L. DARLOW, M. F. WHEELER, AND R. P. KENDALL, *Self-adaptive Galerkin methods for one-dimensional two-phase immiscible flow,* SPE 7679, 5th SPE Symposium on Reservoir Simulation, Denver, 1979.
[41] J. DOUGLAS, JR., T. DUPONT, AND H. H. RACHFORD, JR., *The application of variational methods to waterflooding problems,* J. Canad. Pet. Tech., 8 (1969), pp. 79–85.
[42] J. DOUGLAS, JR., R. E. EWING, AND M. F. WHEELER, *Approximation of the pressure by a mixed method in the simulation of miscible displacement,* RAIRO Anal. Numer., 17 (1983), pp. 17–33.
[43] ———, *A time-discretization procedure for a mixed finite element approximation of miscible displacement in porous media,* RAIRO Anal. Numer., to appear.

[44] J. DOUGLAS, JR., R. P. KENDALL, AND M. F. WHEELER, *Long wave regularization of one-dimensional, two-phase, immiscible flow in porous media*, in Finite Element Methods for Convection-Dominated Flows, AMD Vol. 34, ASME, New York, 1979, pp. 201–211.

[45] J. DOUGLAS, JR., D. W. PEACEMAN, AND H. H. RACHFORD, JR., *A method for calculating multidimensional immiscible displacement*, Trans. AIME, 216 (1959), pp. 297–306.

[46] J. DOUGLAS, JR. AND J. E. ROBERTS, *Numerical methods for a model for compressible miscible displacement in porous media*, to appear.

[47] J. DOUGLAS, JR. AND T. F. RUSSELL, *Numerical methods for convection-dominated diffusion problems based on combining the method of characteristics with finite element or finite difference procedures*, SIAM J. Numer. Anal., 19 (1982), pp. 871–885.

[48] J. DOUGLAS, JR. AND M. F. WHEELER, *Implicit time-dependent variable grid finite-difference methods for the approximation of a linear waterflood*, Math. Comp., 40 (1983), pp. 107–121.

[49] J. DOUGLAS, JR., M. F. WHEELER, B. L. DARLOW, AND R. P. KENDALL, *Self-adaptive finite element simulation of miscible displacement in porous media*, SIAM J. Sci. Statist. Comp., to appear.

[50] R. E. EWING, *Determination of coefficients in reservoir simulation*, in Numerical Treatment of Inverse Problems for Differential and Integral Equations, P. Deuflhardt and E. Hairer, eds., Birkhauser, Boston, 1982.

[51] R. E. EWING AND R. F. HEINEMANN, *Incorporation of mixed finite element methods in compositional simulation for reduction of numerical dispersion*, SPE 12267, SPE 7th Symposium on Reservoir Simulation, San Francisco, Nov. 15–18, 1983.

[52] R. E. EWING AND T. F. RUSSELL, *Efficient time-stepping methods for miscible displacement problems in porous media*, SIAM J. Numer. Anal., 19 (1982), pp. 1–66.

[53] ———, *Multistep Galerkin methods along characteristics for convection-diffusion problems*, in Advances in Computer Methods for Partial Differential Equations IV, R. Vichnevetsky and R. S. Stepleman, eds., IMACS, Rutgers Univ., New Brunswick, NJ, 1981, pp. 28–36.

[54] R. E. EWING, T. F. RUSSELL, AND M. F. WHEELER, *Simulation of miscible displacement using mixed methods and a modified method of characteristics*, SPE 12241, 7th SPE Symposium on Reservoir Simulation, San Francisco, Nov. 15–18, 1983.

[55] R. E. EWING AND M. F. WHEELER, *Galerkin methods for miscible displacement problems in porous media*, SIAM J. Numer. Anal., 17 (1980), pp. 351–365.

[56] ———, *Galerkin methods for miscible displacement problems with point sources and sinks—unit mobility ratio case*, Proc. Special Year in Numerical Anal., Lecture Notes #20, Univ. Maryland, College Park, 1981, pp. 151–174.

[57] ———, *Computational aspects of mixed finite element methods*, in IMACS Transactions on Scientific Computation, vol. 1, R. S. Stepleman, ed., North-Holland, Amsterdam, 1983.

[58] R. S. FALK AND J. E. OSBORN, *Error estimates for mixed methods*, RAIRO Anal. Numer., 14 (1980), pp. 249–277.

[59] A. O. GARDER, JR., D. W. PEACEMAN AND A. L. POZZI, JR., *Numerical calculation of multidimensional miscible displacement by a method of characteristics*, Soc. Pet. Eng. J., 4 (1964), pp. 26–36.

[60] J. GLIMM, E. ISAACSON, D. MARCHESIN, AND O. MCBRYAN, *Front tracking for hyperbolic systems*, Adv. Appl. Math, 2 (1981), pp. 91–119.

[61] J. GLIMM, D. MARCHESIN, AND O. MCBRYAN, *Statistical fluid dynamics: unstable fingers*, Comm. Math. Phys., 74 (1980), pp. 1–13.

[62] ———, *The Buckley–Leverett equation: theory, computation and application*, in Trends in Applications of Pure Mathematics in Mechanics, R. J. Knops, ed., Edinburgh, 1979.

[63] ———, *Unstable fingers in two phase flow*, Comm. Pure Appl. Math., 34 (1981), pp. 53–75.

[64] ———, *A numerical method for two phase flow with an unstable interface*, J. Comp. Phys., 39 (1981), pp. 179–200.

[65] B. HABERMANN, *The efficiency of miscible displacement as a function of mobility ratio,* Trans. AIME, 219 (1960), pp. 264–272.
[66] A. HARTEN, P. D. LAX, AND B. VAN LEER, *On upstream differencing and Godunov-type schemes for hyperbolic conservation laws,* SIAM Rev., 25 (1983), pp. 35–62.
[67] L. J. HAYES, R. P. KENDALL, AND M. F. WHEELER, *The treatment of sources and sinks in steady-state reservoir engineering simulations,* in Advances in Computer Methods for Partial Differential Equations II, R. Vichnevetsky, ed., IMACS, Rutgers Univ., New Brunswick, NJ, 1977, pp. 301–306.
[68] M. K. HWANG, W. R. JINES, AND A. S. ODEH, *An in-situ combustion process simulator,* SPE 9450, 55th Annual SPE Fall Conference, Dallas, 1980.
[69] E. ISAACSON, *Global solution of a Riemann problem for a non-strictly hyperbolic system of conservation laws arising in enhanced recovery,* J. Comp. Phys., to appear.
[70] J. JAFFRE, *Formulation mixte d'écoulements diphasiques incompressibles dans un milieu poreux,* Rapport INRIA 37, INRIA, Le Chesnay, France, 1980.
[71] H. KAZEMI, C. R. VESTAL, AND G. D. SHANK, *An efficient multicomponent numerical simulator,* Soc. Pet. Eng. J., 18 (1978), pp. 355–366.
[72] R. P. KENDALL, G. O. MORRELL, D. W. PEACEMAN, W. J. SILLIMAN, AND J. W. WATTS, *Development of a multiple application reservoir simulator for use on a vector computer,* SPE 11483, SPE Middle East Oil Technical Conference, Manama, Bahrain, 1983, pp. 335–348.
[73] E. J. KOVAL, *A method for predicting the performance of unstable miscible displacement instability,* Soc. Pet. Eng. J., 3 (1963), pp. 145–154.
[74] L. W. LAKE AND G. J. HIRASAKI, *Taylor's dispersion in stratified porous media,* Soc. Pet. Eng. J., 18 (1978), pp. 459–468.
[75] A. LALLEMAND-BARRES AND P. PEAUDECERF, *Investigation of the relationship between the value of the macroscopic dispersiveness of an aquifer medium, its other characteristics, and the measurement conditions—bibliographic study,* Bull. BRGM Ser. 2, §3, #4 (1978), pp. 277–284.
[76] P. LOTSTEDT, *A front-tracking method applied to Burgers' equation and two-phase porous flow,* J. Comp. Phys., 47 (1982), pp. 211–228.
[77] O. MCBRYAN, *Elliptic and hyperbolic interface refinement,* in Boundary Layers and Interior Layers, Computational and Asymptotic Methods, J. J. H. Miller, ed., Boole Press, Dublin, 1980.
[78] ———, *Computing discontinuous flows,* Proc. Meeting on Fronts, Patterns, and Interfaces, Los Alamos, NM, 1983.
[79] ———, *Shock tracking methods in 2-D flows,* Proc. 9th National Applied Mechanics Congress, Cornell Univ., Ithaca, NY, 1982.
[80] C. A. MILLER, *Stability of moving surfaces in fluid systems with heat and mass transport II. Combined effects of transport and density difference between phases,* AIChE J., 19 (1973), pp. 909.
[81] ———, *Stability of moving surfaces in fluid systems with heat and mass transport III. Stability of displacement fronts in porous media,* AIChE J., 21 (1975), pp. 474–479.
[82] M. MUSKAT, *Flow of Homogeneous Fluids Through Porous Media,* McGraw-Hill, New York, 1937.
[83] L. X. NGHIEM, D. K. FONG, AND K. AZIZ, *Compositional modeling with an equation of state,* Soc. Pet. Eng. J., 21 (1981), p. 687.
[84] J. S. NOLAN, *Numerical simulation of compositional phenomena in petroleum reservoirs,* SPE 4274, 3rd Symposium on Numerical Simulation of Reservoir Performance, Houston, 1973.
[85] A. S. ODEH, *Reservoir simulation—what is it?* J. Pet. Tech., (1969), pp. 1383–1388.
[86] ———, *An overview of mathematical modeling of the behavior of hydrocarbon reservoirs,* SIAM Rev., 24 (1982), pp. 263–273.

[87] H. D. OUTMANS, *Nonlinear theory for frontal stability and viscous fingering in porous media,* Soc. Pet. Eng. J., 2 (1962), pp. 165–176.

[88] D. W. PEACEMAN, *Fundamentals of Numerical Reservoir Simulation,* Elsevier, Amsterdam, 1977.

[89] ———, *Interpretation of well-block pressures in numerical reservoir simulation,* SPE 6893, 52nd Annual Fall Technical Conference and Exhibition, Denver, 1977; Soc. Pet. Eng. J., 18 (1978), pp. 183–195.

[90] D. W. PEACEMAN AND H. H. RACHFORD, JR., *Numerical calculation of multidimensional miscible displacement,* Trans. AIME, 225 (1962), pp. 327–339.

[91] D. Y. PENG AND D. B. ROBINSON, *A new two-constant equation of state,* Ind. Eng. Chem. Fundam., 15 (1965), pp. 59–64.

[92] T. K. PERKINS AND O. C. JOHNSON, *A review of diffusion and dispersion in porous media,* Soc. Pet. Eng. J., 3 (1963), pp. 70–84.

[93] T. K. PERKINS, O. C. JOHNSTON, AND R. N. HOFFMAN, *Mechanics of viscous fingering in miscible systems,* Soc. Pet. Eng. J., 5 (1965), pp. 301–317.

[94] R. L. PERRINE, *A unified theory for stable and unstable miscible displacement,* Soc. Pet. Eng. J., 3 (1963), pp. 205–213.

[95] G. A. POPE AND R. C. NELSON, *A chemical flooding compositional simulator,* Soc. Pet. Eng. J., 18 (1978), pp. 339–354.

[96] T. C. POTEMPA, *Finite element methods for convection dominated transport problems,* Ph.D. thesis, Rice Univ., Houston, 1982.

[97] ———, *Three-dimensional simulation of steamflooding with minimal grid orientation,* SPE 11726, 1983 California Regional Meeting, Ventura, CA, 1983.

[98] H. S. PRICE, J. C. CAVENDISH, AND R. A. VARGA, *Numerical methods of higher order accuracy for diffusion-convection equations,* Trans. AIME, 253 (1972), pp. 293–303.

[99] H. S. PRICE AND K. H. COATS, *Direct methods in reservoir simulation,* Trans. AIME, 257 (1974), pp. 295–308.

[100] P. A. RAVIART AND J. M. THOMAS, *A mixed finite element method for 2nd order elliptic problems,* in Mathematical Aspects of the Finite Element Method, Rome 1975, Lecture Notes in Mathematics, Springer-Verlag, Berlin, 1977.

[101] J. E. ROBERTS, *A numerical method for a model for two-phase compressible immiscible displacement in porous media,* RAIRO Anal. Numer. to appear.

[102] G. E. ROBERTSON AND P. T. WOO, *Grid-orientation effects and the use of orthogonal curvilinear coordinates in reservoir simulation,* Soc. Pet. Eng. J., 18 (1978), pp. 13–19.

[103] T. F. RUSSELL, *An incompletely iterated characteristic finite element method for a miscible displacement problem,* Ph.D. thesis, Univ. Chicago, 1980.

[104] ———, *Finite elements with characteristics for two-component incompressible miscible displacement,* SPE 10500, 6th SPE Symposium on Reservoir Simulation, New Orleans, 1982.

[105] ———, *Galerkin time-stepping along characteristics for Burgers' equation,* Proc. 10th IMACS World Congress on System Simulation and Scientific Computation, Montreal, 1982, pp. 43–45; in Scientific Computing, R. Stepleman et al., eds., IMACS, North-Holland, Amsterdam, 1983, pp. 183–192.

[106] P. G. SAFFMAN, *Theory of dispersion in porous media,* J. Fluid Mech., 6 (1959), pp. 321.

[107] P. H. SAMMON, *Numerical approximations for a miscible displacement process in porous media,* SIAM J. Numer. Anal., to appear.

[108] A. E. SCHEIDEGGER, *Physics of Flow Through Porous Media,* 3rd ed., Univ. Toronto Press, Toronto, 1974.

[109] A. SETTARI, H. S. PRICE, AND T. DUPONT, *Development and application of variational methods for simulation of miscible displacement in porous media,* Soc. Pet. Eng. J., 17 (1977), pp. 228–246.

[110] H. G. SPILLETTE, J. G. HILLESTAD, AND H. L. STONE, *A high stability sequential solution*

approach to reservoir simulation, SPE 4542, 48th Annual SPE Fall Meetings, Las Vegas, Sept. 1973.

[111] A. SPIVAK, H. S. PRICE, AND A. SETTARI, Solution of the equation for multidimensional, two-phase immiscible flow by variational methods, Soc. Pet. Eng. J., 17 (1977), pp. 27–41.

[112] R. H. TRIMBLE AND A. E. MCDONALD, A strongly coupled, implicit well coning model, Soc. Pet. Eng. J., 21 (1981), pp. 454–455.

[113] H. J. WELGE, A simplified method for computing oil recovery by gas or water drive, Trans. AIME, 195 (1952), pp. 91–98.

[114] J. A. WHEELER, H. G. WEINSTEIN, AND E. G. WOODS, Numerical model for steam simulation, SPE 4759, 3rd SPE Improved Oil Recovery Symposium, Tulsa, OK, 1974.

[115] M. F. WHEELER, A self-adaptive finite difference procedure for one-dimensional, two-phase, immiscible flow, Seminar on Numerical Analysis and Applications to Continuum Physics, Colecao, Atas 12, Rio de Janeiro, 1980.

[116] M. F. WHEELER AND B. L. DARLOW, Interior penalty Galerkin methods for miscible displacement problems in porous media, in Computational Methods in Nonlinear Mechanics, J. T. Oden, ed., North-Holland, Amsterdam, 1980, pp. 485–506.

[117] M. F. WHEELER, R. E. EWING, R. W. JONES, AND R. FONTECIELLO, Mixed finite element methods for elliptic partial differential equations, Proc. 10th IMACS World Congress on Systems Simulation and Scientific Computation, Montreal, August 8–13, 1982, pp. 40–42.

[118] J. L. YANOSIK AND T. A. MCCRACKEN, A nine-point finite-difference reservoir simulator for realistic prediction of unfavorable mobility ratio displacements, SPE 5734, 4th Symposium on Numerical Simulation of Reservoir Performance, Los Angeles, Feb. 1976; Soc. Pet. Eng. J., 18 (1978), pp. 253–262.

[119] L. C. YOUNG, A finite-element method for reservoir simulation, Soc. Pet. Eng. J., 21 (1981), pp. 115–128.

[120] L. C. YOUNG AND R. E. STEPHENSON, A generalized compositional approach for reservoir simulation, SPE 10516, 6th SPE Symposium on Reservoir Simulation, New Orleans, 1982.

CHAPTER II

Finite Element and Finite Difference Methods for Continuous Flows in Porous Media

THOMAS F. RUSSELL AND MARY FANETT WHEELER

1. Introduction. Miscible displacement is an enhanced oil-recovery process that has attracted considerable attention in the petroleum industry over the last 30 years. It involves injection of a solvent at certain wells in a petroleum reservoir, with the intention of displacing resident oil to other wells for production. This oil may have been left behind after primary production by reservoir pressure and secondary production by waterflooding. The economics of the process can be precarious, because the chemicals it requires are expensive and the performance of the displacement is by no means guaranteed. Complex physical behavior in the reservoir will determine whether enough additional oil is recovered to make the expense worthwhile.

Mathematically, the process is described by a convection-dominated parabolic partial differential equation for each chemical component in the system. These equations are nonlinear and strongly coupled. By summing the component equations, one can obtain an equation that determines the pressure in the system; this nonlinear equation is elliptic or parabolic, according to whether the system is incompressible or compressible. Thus, in this problem one encounters elliptic, parabolic, and near-hyperbolic equations with complicated nonlinear behavior. It is an intricate problem to approximate numerically, and good numerical modeling is critical in the industry for accurate prediction of costly projects.

This paper is concerned with several numerical procedures developed in the last few years for simulation of miscible displacement. This represents work of Douglas N. Arnold, David C. Brown, Bruce L. Darlow, Jim Douglas Jr., Todd Dupont, Richard E. Ewing, Richard P. Kendall, Thomas C. Potempa, Jean E. Roberts, Peter H. Sammon, and the authors. Theoretical and numerical results have been obtained for a wide variety of methods. This research has been mostly directed at a model system of two coupled equations for incompressible two-component single-phase displacement, but the techniques can be extended to more complex systems representative of realistic reservoir flow. In particular, we believe that similar procedures are appropriate for immiscible as well as miscible displacement, and therefore for most problems of reservoir simulation. In the model problem, the dependent variable of most physical interest is the concentration $c(x, t)$, $0 \leq c \leq 1$, of the solvent in the fluid mixture. Figures 1 through 3 give a qualitative idea of the time evolution of level curves of c in a typical stable displacement. The displacement is on an areal five-spot pattern as in Fig. 3 of Chapter I.

The paper is organized as follows. Section 2 discusses the physics of miscible and immiscible displacement at some length in order to justify our inclusion of physical diffusion-like terms in our numerical procedures. There has been some sentiment in the mathematical literature that these displacement processes possess so little parabolic character that they should be modeled as discontinuous hyperbolic flows; we have taken a different view, and we present supporting reasoning here. In the next section we derive the partial differential equations that represent the chosen physical model. Section 4 surveys the usual time-stepping methods considered in the petroleum industry for such problems and indicates why we decided to use sequential procedures. As a result, we treat the two model equations, for pressure and concentration, with different sets of methods described in §§5 and 6. Before presenting these methods we note the procedures that have been used in the industry and point out some of their inadequacies. Methods for the pressure equation include standard and mixed finite elements with near-well logarithmic singularities removed or not. We demonstrate first-order convergence of the commonly used inconsistent block-centered finite difference method by showing that it is equivalent to a certain mixed method with a particular quadrature rule. For concentration, we introduce interior-penalty Galerkin procedures on continuous and discontinuous spaces, a modified method of characteristics using finite differences or finite elements, and a finite difference method using cell balances based on finite element concepts.

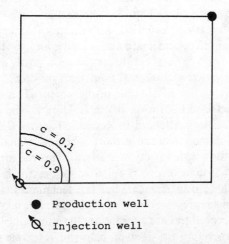

FIG. 1. *Level curves of solvent concentration in the early stages of miscible displacement.*

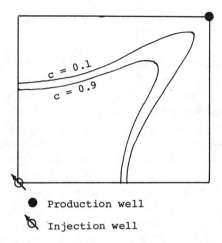

FIG. 2. *Level curves of solvent concentration approaching solvent breakthrough.*

Convergence theorems are cited for all procedures for both equations, together with numerical computations for the new methods. Section 7 shows how the pressure and concentration schemes have been tied together to solve the coupled system of equations, again presenting theoretical and numerical results. Section 8 indicates ideas and directions for future research, and the final section gives conclusions and the authors' opinions about the results obtained so far.

This paper is primarily expository. Our principal objective is to show what has been done and why it has been done, and to give the reader a feel for the results. Accordingly, we shall try to keep technical details to a minimum and to avoid the

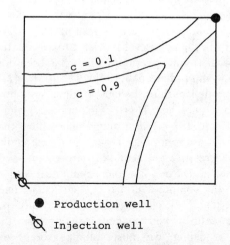

FIG. 3. *Level curves of solvent concentration after solvent breakthrough.*

lengthy proofs that pervade numerical analysis of coupled nonlinear systems. References for further details are indicated in the text.

2. Discussion of physical processes. As noted in Chapter I of this volume, reservoir simulation must begin with a physical model that adequately describes the significant flow phenomena. The nature of the physical model will affect the appropriateness of various possible mathematical and numerical models. Our mathematical models for the displacement processes considered in this chapter are partial differential equations of convection-diffusion type, whose solutions exhibit moving fronts that are sharp but continuous. The corresponding numerical methods we shall discuss are based on finite differences and finite elements, which, in their standard forms, treat diffusive processes effectively but have difficulty with convection. We shall present modifications of these standard forms designed to handle convection better without losing the ability to simulate diffusion rigorously. This section sets the background for these mathematical and numerical models by surveying the engineering literature to find out what processes should be considered in a physical model.

It was pointed out in §3 of Chapter I that compositional, chemical, and thermal simulation combine aspects of miscible and immiscible displacement when viewed mathematically. Since our objective is to test new numerical ideas at a research level, we confine our discussion to miscible and immiscible displacement. We emphasize miscible displacement because most of our work has been in that direction, but we shall indicate why we believe that analogous ideas make sense for immiscible displacement.

Miscible and immiscible displacement both involve convection, or physical transport, of fluids through the porous medium. At a macroscopic level, this process is governed by Darcy's law, (3.1) and (3.22) of Chapter I. We must stress the word "macroscopic"; at the microscopic level of a pore size, convection is highly irregular. In fact, under certain assumptions, Darcy's law has been derived rigorously from the pore-level Navier–Stokes equations by a homogenization process of local volume averaging. This has been done for single-phase [89], [91], [113] and multiphase flow [90], [92], [65], [42]. As emphasized in [113], the characteristic length of the local volumes used for averaging must be much greater than a pore length (10^{-4} meters, say) and much less than a reservoir length (10^2 to 10^3 m). For single-phase flow, the permeability and pressure in Darcy's law are volume-averaged quantities; similarly, for multiphase flow, saturations and relative permeabilities are volume-averaged. This means that in a model based on Darcy's law, one cannot deal with variations on a finer scale than that of the volume averages; in particular, one cannot deal with discontinuities. Claridge [15], citing geological literature, points out that rocks have characteristic scales of heterogeneity ranging from the pore level to the reservoir level; in modeling, we must volume average over heterogeneities significantly smaller than grid size, so our averaging volumes must be of the order of the grid size. Thus, microscopic heterogeneities are lost when we pass to Darcy's law; the next scale of heterogeneity must be averaged in modeling, and we shall see that this can be modeled by dispersion; heterogeneities at or larger than grid size can be modeled directly with equation coefficients. If reservoir flow processes are in fact discontinuous (we contend below that they are not), they must not be modeled by fundamentally macroscopic principles. These observations show that flow in a porous medium is to be distinguished from other phenomena of fluid dynamics.

Miscible displacement is also characterized by the processes of diffusion and dispersion, about which there is considerable literature. Diffusion of one fluid into another, which is the result of random motion of molecules, is governed by the Fick equation [71]

$$(2.1) \qquad \frac{\partial V}{\partial t} = -d_m A \frac{\partial C}{\partial x},$$

where V is the volume of one fluid on the positive-x side of a plane of cross-sectional area A normal to the x-direction, C is the concentration of that fluid, and d_m is the molecular diffusion coefficient (units: length^2/time) in a nonporous medium. In a porous medium, d_m is multiplied by $1/F\phi$, where F is the electrical resistivity factor of the formation and ϕ is the porosity [71]. F can usually be roughly approximated by ϕ^2 [36], so that the porous coefficient becomes ϕd_m as in (3.33) of Chapter I. This coefficient is typically of the order of 10^{-5} cm^2/sec or 10^{-3} ft^2/day [45], which is quite small.

Most of our discussion will deal with dispersion, a term that covers assorted physical phenomena. At a microscopic level, it is mechanical mixing caused by flow through tiny capillaries. In a single capillary tube, the theory of Taylor [99] leads to a diffusion-like term whose coefficient is proportional to the square of fluid velocity. This would extend to a bundle of straight capillary tubes. In a pack of sand grains, Brigham et al. [10] observe that a process that they call "eddy diffusion" will take place, in which flow channels will separate and meet again, with fluids being dispersed by virtue of traveling different distances. There the amount of dispersion should depend on the distance traveled by the fluids and not on their velocity; since a diffusion-like coefficient represents fluid movement per unit time and the time corresponding to a given distance will be inversely proportional to velocity, the coefficient must be proportional to velocity. Laboratory measurements with reservoir cores (rock samples) have found coefficients proportional to velocity to approximately the power 1.2 [10], [71]; this has been explained by the core combining the properties of sand packs and capillary tubes [10] or by diffusion failing to equalize concentration within single pores at high velocities [71].

Perhaps more significant for our purposes is dispersion at a macroscopic level. This is caused by macroscopic variations in rock permeability and porosity, and it involves differences in velocities of individual fluid elements as they move through the rock [106]. Because of the qualitative similarity between this process and microscopic mechanical mixing, it is not surprising that a distributed (log-normal) permeability system behaves as if macroscopic dispersion could be represented by a diffusion-like term [106] with coefficient proportional to velocity. A specific process of this type, observed in [57] and called Taylor's dispersion, concerns flow at different rates in adjacent horizontal rock layers combined with vertical dispersion between the layers, resulting in a large effective horizontal dispersion; a similar process had been studied by Taylor [99], [100]. Because macroscopic dispersion is most important for us, we shall use the exponent 1 instead of 1.2 in our work.

In general, we should note that mechanical mixing and, by analogy, macroscopic dispersion have two coefficients, one parallel (longitudinal) to flow and the other normal (transverse); the longitudinal coefficient may be 30 times the transverse one [71], [57]. They are commonly denoted in the literature by α_l, α_t (units: length) and multiplied by the interstitial velocity v (the velocity of fluid movement; u/ϕ, if u is the Darcy velocity) to obtain a diffusion-like coefficient. This coefficient is then substituted for D/ϕ in an equation like that obtained from (3.32) of Chapter I divided through by ϕ; thus, α_l and α_t are equivalent to d_l and d_t as defined in (3.33) of Chapter I.

We indicate next the potential importance of dispersion in applications of miscible displacement. The overriding significance of viscous fingering was enunciated in Chapter I, and the behavior of viscous fingers is primarily controlled by mixing [73]. Specifically, transverse dispersion can enhance the effectiveness of miscible displacement by suppressing fingers altogether [103], [73], [15], or it can influence the geometry of fingers [96]. Longitudinal dispersion blunts the tips of fingers [44], affecting their growth patterns. In a linear system, the minimum finger width that will not be damped out is proportional to $\sqrt{ex\alpha_t}$, where e is the displacement efficiency (ratio of injected fluid to total fluid behind the leading edge of the front; equal to 1 for a piston-like displacement) and x is the distance traveled; a similar relationship holds in a radial system [73]. The onset of fingers is also related to dispersion, as in the theories of Perrine [74], later supported by experiments [75], and Heller [48], in which a front was stable below a critical velocity proportional to $1/\alpha_t$.

Dispersion can likewise be harmful to oil recovery by miscible displacement. A typical miscible process involves injection of a solvent slug that is miscible with oil provided that its concentration exceeds a minimum that is a function of pressure [52], [44], [115]. The slug will be followed by another material so that the slug, which is expensive, may be recovered; examples include propane followed by gas, alcohol followed by water [52], and a micellar solution followed by polymer water [115], [39]. Dispersion can dissipate the slug below minimum miscibility concentration, rendering the process immiscible [71], [44], [115]. As discussed in Chapter I, this results in much residual oil being left trapped in small pores by interfacial tension.

The relationship between dispersion and ion exchange is extremely important for chemical flooding. Recovery efficiency of a chemical slug depends on phase behavior, mobility, and interfacial effects, all of which depend in turn on ionic environment [56]. Qualitative differences in ion concentrations are shown in [56] depending on whether dispersion is incorporated in simulation or not, and a dispersion-induced exchange process is described in which two fluids of similar ionic properties mix to create different properties. With laboratory experiments, Gupta [43] shows that the behavior of a field micellar (sulfonate/polymer) flood cannot be explained without significant dispersion. Sulfonate partitions into the oil phase because of ion exchange, while polymer stays in the water phase. Since the dispersion coefficient for sulfonate in the oil phase is an order of magnitude larger than water-phase coefficients, sulfonate breaks through at the production well before a nondispersive model would predict. Thus, sulfonate is produced earlier than polymer, and the entire process requires more sulfonate than would be the case without dispersion.

If dispersion is potentially important, as we have seen, the question remains of whether it is significant, so that its effects are felt in practical problems. As we noted, molecular diffusion is quite small; if dispersion were no more important than diffusion, its effects would be of little interest. In the late fifties and early sixties, when there was a flood of interest in miscible recovery processes, some investigators took that view. In these experiments, linear tubes were packed uniformly with unconsolidated sand [6] or glass powder [103], and a two-dimensional quarter five-spot pattern (Fig. 3 of Chapter I) was packed uniformly with glass beads [55]. The porous media of petroleum reservoirs, however, are consolidated rocks, and linear [45], [96] and five-spot [44] experiments in consolidated media led to the opposite conclusion that mechanical mixing (dispersion) dominates diffusion. This conclusion was based on the observation that the outflowing concentration profile after a flood through a consolidated medium was independent of the flow rate (velocity); as we noted above, this implies a mechanical mixing model with dispersion coefficient proportional to velocity. Indeed, the experiments of Brigham et al. [10] found that the microscopic heterogeneities of consolidated rock will increase the dispersion coefficient by two orders of magnitude over its value for glass beads; this supports the findings of Habermann [44], in which the mixing zone was 8 to 20 percent of the area of the quarter five-spot, as opposed to those of Lacey et al. [55], where it was 1.6 to 5 percent. Discussions published with [55] indicated serious doubts of many readers as to whether dispersion was properly scaled from the laboratory to field reservoirs in that study. The work of Koch and Slobod [52] with an unconsolidated-sand tube much longer (123 feet) than those of the other experiments concluded that dispersion was the predominant mixing mechanism.

Thus, it appears to us that dispersion, even at the microscopic level, is vastly more significant than molecular diffusion alone. This viewpoint is advanced in relation to viscous fingering in [45] and [44], where it is noted that fingering increases the length of the interface by making it irregular, thereby increasing the effect of dispersion. Handy [45] finds that dispersion dominates diffusion even in this case, where dispersion acts mainly through the transverse α_t, the smaller of its two coefficients.

Additionally, we have the phenomenon of macroscopic dispersion, which came to be discussed in more recent literature. It has been pointed out [106], [58] that the scale of macroscopic dispersion is related to the scale of heterogeneities in the reservoir, so that small laboratory models are inadequate to measure this behavior and will therefore understate the dispersion that will occur in the field. Claridge [15] echoes this and identifies how heterogeneities and dispersion interact in the all-important process of viscous fingering. Even without large-scale heterogeneities, he contends that a reservoir-scale displacement will show a mixed zone of 5 to 10 percent of the total system length; large-scale heterogeneities would increase this.

Field measurements of macroscopic dispersion are hard to come by, since they are complicated by nonlinear flow, heterogeneities, and operational difficulties in the field [57]. A two-well test, which is the type of interest to us, is cited in [57] leading to an estimate of 8 feet for α_l. As we shall see in §3, the Peclet number (ratio of dimensionless convection and dispersion coefficients) of a miscible displacement is L/α_l, if L is a characteristic reservoir length and diffusion is neglected; in this case it is of the order of 100 to 300. A tracer test in the same field considered in [43] was approximately matched by $\alpha_l = 3$ feet [116]; the maximum produced slug concentration was actually a bit higher in the simulation than in the field, indicating that the actual value of α_l might be slightly greater. An extensive review of field dispersion calculations for pollutant transport in aquifers has been published in France [58]. This article compiled tracer measurements from 27 sources and related dispersion coefficients to distance traveled and flow velocity. The authors found that velocity was not important, supporting the dispersion model [106] of macroscopic mixing, but that distance was important. Longer distance correlated with higher dispersivity. Lallemand-Barres et al. postulated that a critical distance existed beyond which this correlation would no longer hold (presumably beyond the largest scale of heterogeneity), but that most of their cited measurements did not reach that distance. For a sizable majority of their cases, the Peclet number was between 10 and 30. The theory of [106] leads to a Peclet number of approximately $2/S^2$ for a log-normal permeability distribution, where the variance S^2 is roughly the difference of the arithmetic and harmonic average permeabilities divided by the harmonic average; in a field example, S^2 was found to be about 0.1, for a Peclet number of about 20. These similar results from widely varying sources indicate to us that convection-diffusion models of this range of Peclet number are appropriate for simulation of miscible displacement. Even without dispersion, research with graded-viscosity slugs [95], [54] has demonstrated reduced viscous fingering for field application, forcing one to consider modeling of continuous fronts. But the views we can find in current literature [57], [116], [115] do not support the contention that molecular diffusion is the dominant mixing mechanism and dispersion is negligible.

It could still be questioned whether conventional convection-diffusion-type equations properly average the physical behavior over all scales of heterogeneity. The theoretical work of Warren and Skiba [106] led to this type of model, as did the laboratory study of Yellig and Baker [115]. In the latter study, dispersive mixing in short heterogeneous systems was found to appear homogeneous in longer systems, so that an effective dispersion coefficient could model it. This effective coefficient was found to increase with system heterogeneity or system length. This modeling concept is supported by Whitaker [112] and Gray [41], in which local volume averaging of the dispersion equation was carried out, leading to an equation in volume-averaged variables that has the same form as the standard convection-diffusion equation. The dispersion tensor contains terms proportional to velocity (mechanical mixing model of Brigham et al. [10]) and velocity squared (capillary-tube model of Taylor [99]).

In our work, we can model heterogeneities at or larger than mesh scale directly with variable coefficients in the partial differential equations. This will allow detailed study of such phenomena as macroscopic viscous fingering at mesh scale. Scales between the Darcy volume averages and the mesh will be modeled with the dispersion concepts justified by this discussion.

We now turn our attention to immiscible displacement and outline our reasons for advocating similar modeling techniques for it. Comparing (3.29b) and (3.32b) of Chapter I, we see that the immiscible analogue of dispersion is capillary pressure. This pressure difference across a phase interface within a pore is a function of the interfacial tension and the curvature of the interface. We focus now on the importance of interfacial tension in the physics of immiscible displacement.

An oil globule trapped in a pore is held there by interfacial tension. There is evidence [47], [93] that when such a globule moves, it does so in small discrete jumps that occur when the local pressure gradient overcomes the interfacial tension and collapses the interface. Between these jumps, the interface slowly deforms. To recover more oil by immiscible displacement, the quotient of pressure gradient and interfacial tension must exceed a critical value; once this happens, further increases in this quotient result in significantly greater oil recovery [104], [98], [93]. The ability to recover more oil means that the residual oil saturation, which is the maximum oil saturation for which the oil relative permeability (see (3.23) of Chapter I) is zero, has been decreased with decreasing interfacial tension. More generally, the entire relative-permeability function, on which the efficiency of immiscible displacement depends very strongly, is affected by interfacial tension, both for steady [59], [2] and unsteady [105], [70], [2] flow. These irreversible forces, as opposed to convection, are responsible for the hysteresis of capillary pressure and relative permeability [47] (the functions differ depending on whether the wetting phase displaces the nonwetting phase or vice versa). All of these considerations govern the microscopic efficiency of immiscible displacement.

The macroscopic efficiency of immiscible displacement is governed by its stability or lack thereof (viscous fingering). Theories of the onset of instability have found it to occur above a critical flow velocity with wavelengths (finger widths) above a cutoff dependent on interfacial tension [83], [13], [64]. Peters and Flock [76] developed a dimensionless parameter, inversely proportional to interfacial tension, such that instability develops if the parameter exceeds a critical value. A difficult issue here is that interfacial tension acts at the microscopic pore level, while macroscopic fingering demands a macroscopic concept; probably the way to resolve this is through an effective macroscopic interfacial tension [40], justified by experiments displaying a cutoff wavelength. This may possibly be best understood in terms of volume averaging that would take account of the physical interfacial tension [94]. As in the miscible case, this would preclude modeling that dealt with fronts of sharpness beyond grid scale.

Another important development is the increasing view that miscible and immiscible displacement share many characteristics. Perkins and Johnston [72] conducted laboratory experiments that led to a concept of "immiscible dispersion." When water was injected into a bead-packed model saturated with oil and irreducible water, small fingers initiated, broke into a tree structure, and then deteriorated into a graded saturation zone. Side-by-side injection of water and oil into linear consolidated bead models gave rise to a transverse two-phase zone with saturation profile analogous to the concentration profile of miscible displacement; the dispersion coefficient was proportional to velocity, as in the miscible case. It was theorized that this process, like miscible dispersion, was caused by stream splitting during convection. It would seem that the macroscopic dispersion concepts should carry over to immiscible displacement as well, especially since field tests such as those that give large dispersion coefficients [43], [57], [58] are really combinations of miscible and immiscible flow regimes.

For both miscible and immiscible displacement, it seems to us that the complex physical behavior is largely a function of the mathematical regularizers—dispersion and interfacial tension. These processes should not be neglected in modeling; they, and how they interact with rock heterogeneities, appear to be precisely what needs to be understood. By using methods based on finite differences and finite elements, which handle these processes well, and by incorporating heterogeneities (at scales greater than volume averages) into the dispersion and permeability coefficients of the differential equations, we hope to model the physics realistically and rigorously. In particular, we can demonstrate theorems that tell us that our procedures converge to the solutions of the differential equations chosen as models. We feel that procedures based on propagation of discontinuities, while leading to some excellent mathematical and numerical work with application to other fields, are not appropriate for the problems of petroleum reservoir simulation.

3. Derivation of differential system for miscible displacement. Having discussed the physical phenomena that should be included, we are in a position to derive partial differential equations describing miscible displacement. Many of the necessary ideas are the same as those given in Chapter I for single-phase flow and two-phase immiscible flow. In particular, Darcy's law and conservation of mass are fundamental. Here, however, we have two components (denote them by subscripts o and s for oil and solvent) flowing together as one phase, so we use conservation of mass for each component instead of the phase. In addition, we must deal with fluid fluxes due to diffusion and dispersion of the two components into each other. An equation of state relating density to pressure is also important in this context, but for simplicity we shall assume that the fluids are incompressible.

As in Chapter I, consider a volume element V. Recalling the volume averaging discussed in §2, V should be at least as large as the scale of the averaging volumes. Using the notation of Chapter I, the rate of mass accumulation of component i in V is

$$(3.1) \quad \frac{d}{dt} \int_V \phi \rho c_i(x, t) dx, \qquad i = o, s,$$

where $c_i \in [0, 1]$ is the concentration of component i in the single phase. Darcy's law ((3.1) of Chapter I) yields the volumetric convective flow rate of the mixture per unit cross-sectional area; restricting this to component i, we obtain

$$(3.2) \quad \tilde{u}_i = -c_i \frac{k}{\mu}(\nabla p - \rho g \nabla z) = c_i \tilde{u}.$$

From Fick's law (2.1), as modified in §2 for porous media, we obtain the diffusive flow rate

$$(3.3) \quad \tilde{u}_{i,\text{diff}} = -\phi d_m \nabla c_i.$$

In §2, we decided to use the first-power law for dispersion, giving rise to the dispersive flow rate

$$(3.4) \quad \tilde{u}_{i,\text{disp}} = -d_l |\tilde{u}| \frac{\partial c_i}{\partial \hat{e}_l} \hat{e}_l - d_t |\tilde{u}| \frac{\partial c_i}{\partial \hat{e}_t} \hat{e}_t,$$

where $\hat{e}_l$ and $\hat{e}_t$ are unit vectors in the directions parallel and perpendicular to $\tilde{u}$ (in three dimensions, a third term is added with coefficient d_t). If θ is the angle between $\hat{e}_l$ and the cartesian unit vector $\hat{e}_x$, we use the relations

$$\hat{e}_l = \hat{e}_x \cos\theta + \hat{e}_y \sin\theta,$$
$$\hat{e}_t = -\hat{e}_x \sin\theta + \hat{e}_y \cos\theta,$$
$$\cos\theta = \tilde{u}_x/|\tilde{u}|,$$
$$\sin\theta = \tilde{u}_y/|\tilde{u}|,$$

to rotate coordinates and express (3.4) as

$$\begin{aligned}
\tilde{u}_{i,\text{disp}} = & - d_l|\tilde{u}|\left(\frac{\tilde{u}_x}{|\tilde{u}|}\frac{\partial c_i}{\partial x} + \frac{\tilde{u}_y}{|\tilde{u}|}\frac{\partial c_i}{\partial y}\right)\left(\frac{\tilde{u}_x}{|\tilde{u}|}\hat{e}_x + \frac{\tilde{u}_y}{|\tilde{u}|}\hat{e}_y\right) \\
& - d_t|\tilde{u}|\left(-\frac{\tilde{u}_y}{|\tilde{u}|}\frac{\partial c_i}{\partial x} + \frac{\tilde{u}_x}{|\tilde{u}|}\frac{\partial c_i}{\partial y}\right)\left(-\frac{\tilde{u}_y}{|\tilde{u}|}\hat{e}_x + \frac{\tilde{u}_x}{|\tilde{u}|}\hat{e}_y\right) \\
= & -\frac{d_l}{|\tilde{u}|}\left[\left(\tilde{u}_x^2\frac{\partial c_i}{\partial x} + \tilde{u}_x\tilde{u}_y\frac{\partial c_i}{\partial y}\right)\hat{e}_x + \left(\tilde{u}_x\tilde{u}_y\frac{\partial c_i}{\partial x} + \tilde{u}_y^2\frac{\partial c_i}{\partial y}\right)\hat{e}_y\right] \\
& -\frac{d_t}{|\tilde{u}|}\left[\left(\tilde{u}_y^2\frac{\partial c_i}{\partial x} - \tilde{u}_x\tilde{u}_y\frac{\partial c_i}{\partial y}\right)\hat{e}_x + \left(-\tilde{u}_x\tilde{u}_y\frac{\partial c_i}{\partial x} + \tilde{u}_x^2\frac{\partial c_i}{\partial y}\right)\hat{e}_y\right].
\end{aligned}$$

(3.5)

Note that a pth-power law (e.g., $p = 1.2$) would lead to $|\tilde{u}|^{2-p}$ in the denominator of (3.5). Now we combine (3.1), (3.2), (3.3), and (3.5) with the divergence theorem as in Chapter I, equating accumulation to influx and sources:

(3.6)
$$\int_V \frac{\partial}{\partial t}(\phi\rho c_i)\, dx = -\int_V \nabla\cdot(\rho c_i \tilde{u} + \rho\tilde{u}_{i,\text{diff}} + \rho\tilde{u}_{i,\text{disp}})\, dx + \int_V \tilde{c}_i \rho q\, dx,$$

where $\tilde{c}_i$ is the specified concentration of component i at an injection well and the resident concentration at a producer, and q is the volumetric flow rate per unit volume. We assume incompressibility of the fluids and the rock, and by using (3.5) we can rewrite (3.6) in the form

(3.7) $$\int_V \left[\phi\frac{\partial c_i}{\partial t} - \nabla\cdot(D\nabla c_i - \tilde{u}c_i)\right] dx = \int_V \tilde{c}_i q\, dx, \qquad i = o, s,$$

where D is a diffusion-disperson tensor given by

(3.8) $$D = \phi d_m I + \frac{d_l}{|\tilde{u}|}\begin{pmatrix}\tilde{u}_x^2 & \tilde{u}_x\tilde{u}_y \\ \tilde{u}_x\tilde{u}_y & \tilde{u}_y^2\end{pmatrix} + \frac{d_t}{|\tilde{u}|}\begin{pmatrix}\tilde{u}_y^2 & -\tilde{u}_x\tilde{u}_y \\ -\tilde{u}_x\tilde{u}_y & \tilde{u}_x^2\end{pmatrix}.$$

This tensor is also derived in Peaceman [66]. In three dimensions, one can show that the analogous tensor is

$$\phi d_m I + d_l|\tilde{u}|T + d_t|\tilde{u}|(I - T),$$

where

$$T = \frac{1}{|\tilde{u}|^2} \begin{pmatrix} \tilde{u}_x^2 & \tilde{u}_x\tilde{u}_y & \tilde{u}_x\tilde{u}_z \\ \tilde{u}_x\tilde{u}_y & \tilde{u}_y^2 & \tilde{u}_y\tilde{u}_z \\ \tilde{u}_x\tilde{u}_z & \tilde{u}_y\tilde{u}_z & \tilde{u}_z^2 \end{pmatrix}.$$

We would like to drop the integrations in (3.7) by collapsing V to an arbitrary point, but this requires justification since V is at least on the scale of the averaging volume. The work of Whitaker [112] and Gray [41] demonstrated that, in terms of volume-averaged variables, one can arrive at the same equation we would obtain if we dropped integrations in (3.7). Hence, we have the coupled system

$$(3.9) \qquad \phi \frac{\partial c_i}{\partial t} - \nabla \cdot (D\nabla c_i - \tilde{u}c_i) = \tilde{c}_i q, \qquad i = o, s.$$

The dependent variables are $\tilde{u}$, c_o, and c_s, with the additional constraint of $c_o + c_s = 1$. The two equations can be called an *oil equation* and a *solvent equation*.

A common manipulation with this system is to replace one of the equations (say the oil equation) by the sum of the two. When the equations are added together, the relation $c_o + c_s = 1$ causes the accumulation and diffusion-dispersion terms to drop out, leaving the system

$$(3.10a) \qquad -\nabla \cdot \left(\frac{k}{\mu(c)} (\nabla p - \rho g \nabla z) \right) \equiv \nabla \cdot \tilde{u} = q,$$

$$(3.10b) \qquad \phi \frac{\partial c}{\partial t} - \nabla \cdot (D(\tilde{u})\nabla c - \tilde{u}c) = \tilde{c}q,$$

where $c = c_s$, $\tilde{c} = \tilde{c}_s$. The first equation (3.10a) now represents conservation of mass for the total fluid, the second for the solvent component. The first equation is frequently called the *pressure equation* because it is an elliptic equation for pressure with a concentration-dependent coefficient. The second equation may be considered a *concentration equation* because it is a (usually) convection-dominated parabolic equation for concentration with coefficients depending on pressure through the Darcy velocity. Because (3.10b) depends on p only through $\tilde{u}$, we may also think of (3.10a) as a *velocity equation*.

In (3.10a) the behavior of μ as a function of c is very important to viscous fingering, displacement efficiency, and ultimate oil recovery. This behavior depends on the *mobility ratio* $\mu_o/\mu_s = \mu(0)/\mu(1) = M$. If $M > 1$, the displacement is said to have *adverse mobility ratio* and fingering is expected.

This derivation was not restricted to two components. We could have obtained a system of n equations in (3.9) for an n-component system flowing as one phase, and then used $c_1 + c_2 + \cdots + c_n = 1$ to replace one of the n equations with a pressure equation like (3.10a). By also considering multiphase flow, we could derive a compositional system like (3.34) in Chapter I, with the $\tilde{u}$'s of that equation containing diffusive-dispersive terms as well as Darcy terms. Within each phase, flow would be similar to that described by (3.10), but the phase volume fraction (saturation) would multiply all the terms and change with space and time, and there would be mass transfer between phases. Likewise, the derivation extends readily to compressible flow, as in Peaceman [67], rendering the pressure equation parabolic as noted in Chapter I. For simplicity in trying new methods, we shall confine our attention primarily to the system (3.10), keeping in mind that we want ideas that will extend conveniently to more complex problems in three dimensions.

The system (3.10) is taken to hold over a bounded domain Ω, representing a reservoir or part of a reservoir, through a time interval $J = [0, T]$. All of our computational work to date has dealt with $\Omega \subset \mathbb{R}^2$, but we have theoretical results in $\mathbb{R}^3$ and eventually expect to perform three-dimensional computations with these procedures. The system requires boundary and initial conditions; in reservoir modeling, the usual boundary conditions are no-flow, representing an axis of symmetry or an impermeable boundary,

(3.11a) $\qquad \tilde{u} \cdot \nu = 0, \quad x \in \partial\Omega, \quad t \in J,$

(3.11b) $\qquad (D\nabla c) \cdot \nu - c(\tilde{u} \cdot \nu) = 0, \quad x \in \partial\Omega, \quad t \in J,$

or constant-pressure, applicable when the reservoir is repressurized at the boundary during depletion, say by an aquifer:

(3.12a) $\qquad p(x, t) \equiv p_o, \quad x \in \partial\Omega, \quad t \in J,$

(3.12b) $\qquad$ appropriate condition for c.

In our work, we shall use no-flow conditions, which are the most common in practice. Note that (3.10)–(3.11) determine p only up to an additive constant and that the compatibility condition

$$\int_\Omega q\, dx = \int_\Omega \nabla \cdot \tilde{u}\, dx = \int_{\partial\Omega} \tilde{u} \cdot \nu\, ds = 0,$$

which simply says that the total fluid input to an incompressible system with no-flow boundary must be zero, is forced. Finally, an initial condition

(3.13) $\qquad c(x, 0) = c_0(x), \quad x \in \Omega,$

is required, and initial values for p are determined by (3.10a) and (3.13). If the system were compressible, we would need an initial condition for p as well.

The system (3.10)–(3.11)–(3.13) is in *divergence form,* and for certain numerical procedures we want to consider an alternative *nondivergence form.* Differentiating the convection term in (3.10b) with the product rule and applying

(3.10a), we see that

$$\nabla \cdot (\tilde{u}c) = \tilde{u} \cdot \nabla c + c(\nabla \cdot \tilde{u}) = \tilde{u} \cdot \nabla c + cq,$$

so that (3.10) may be rewritten

(3.14a) $\qquad \nabla \cdot \tilde{u} = q,$

(3.14b) $\qquad \phi \dfrac{\partial c}{\partial t} - \nabla \cdot (D(\tilde{u})\nabla c) + \tilde{u} \cdot \nabla c = (\tilde{c} - c)q.$

Note that, except at the flowing wells where q is nonzero, (3.14) is the same as (3.10). If a numerical approximation of $\tilde{u}$ is divergence-free away from wells (for example, the mixed methods described in §5), the same equivalence will hold numerically.

If, instead of (3.11), we use nonhomogeneous Neumann boundary conditions

$$\tilde{u} \cdot \nu = f, \qquad (D\nabla c) \cdot \nu - c(\tilde{u} \cdot \nu) = g,$$

then we see that the concentration condition can be written

(3.15) $\qquad (D\nabla c) \cdot \nu = fc + g.$

When a Galerkin procedure is applied to (3.14b), (3.15) arises naturally from Green's theorem. Theoretical treatment of this nonlinear boundary condition was carried out in [31], and we shall discuss the issue no further in this paper.

So far we have said little about the source terms of the differential system. In our computational work, we have used sums of Dirac delta functions, representing point sources and sinks. A moment's thought makes one realize that the pressure at a flowing well will then be $+\infty$ or $-\infty$, according as the well is an injector or a producer. Needless to say, infinite or negative pressures make little sense to a petroleum engineer. The true physical pressure around a well will resemble a truncated logarithmic singularity, as in (3.19) of Chapter I. But for the purpose of testing numerical methods in an incompressible regime, it makes sense to see how well they can handle full singularities, since the real problems are almost singular. Once we move to compressible flow, with pressure-dependent physical properties, something else will be done. For similar reasons, we are led to theoretical questions about regularity of singular solutions and convergence rates of numerical schemes to singular solutions.

Finally, we observe the degree to which (3.10b) or (3.14b) is convection-dominated, as defined by its Peclet number. If we neglect molecular diffusion by comparison to dispersion, and if we scale by a characteristic length L (usually the distance between corners for a five-spot pattern as in Fig. 3 of Chapter I), a one-dimensional analogue of (3.14b) becomes

$$\phi \frac{\partial c}{\partial t} - \frac{d_l |\tilde{u}|}{L^2} \frac{\partial^2 c}{\partial x^2} + \frac{\tilde{u}}{L} \frac{\partial c}{\partial x} = (\tilde{c} - c)q.$$

The scaled convection/dispersion ratio L/d_l is the (longitudinal) Peclet number. In two dimensions, a transverse Peclet number L/d_t also applies. In §2 we observed that practical problems should have Peclet numbers in the tens or hundreds, and this is the range considered in our computational work.

4. Possible time-stepping approaches. The system (3.10) or (3.14) is amenable to many possible time-stepping ideas. These fall into three major classes, as discussed in Chapter I: IMPES (implicit pressure, explicit saturation—here concentration plays the role of saturation), fully implicit, and sequential. All three are in common use in the petroleum industry. Our methods are sequential. In this section we give some details of the three alternatives and indicate why we chose to pursue sequential methods.

The IMPES method is applied to a system like (3.10) or (3.14) containing a pressure equation. In passing from time level n to $n + 1$, one first solves (3.10a) for p^{n+1}, lagging all concentration dependencies to the old level c^n. If the system were compressible, with parabolic (3.10a), the time derivative of pressure would be approximated by the backward difference quotient $(p^{n+1} - p^n)/\Delta t$; hence the notion of implicit pressure. The density ρ would be pressure-dependent and would multiply the terms of the equation as in (3.5), and in practical problems the viscosity μ is also pressure-dependent. Thus, the IMPES pressure equation encountered in industry simulation is, unlike (3.10a), nonlinear.

Having obtained p^{n+1}, one then solves explicitly for c^{n+1}. We shall illustrate this in terms of finite differences in one dimension, using the block-centered scheme most popular in the industry. Let $i - 1, i, i + 1$ index three blocks of lengths $\Delta x_{i-1}, \Delta x_i, \Delta x_{i+1}$ with centers located at x_{i-1}, x_i, x_{i+1}. Define the distance between centers, e.g., $\Delta x_{i+1/2} = \frac{1}{2}(\Delta x_i + \Delta x_{i+1})$, and let $x_{i-1/2}, x_{i+1/2}$ be the interface coordinates (see Fig. 4). Now approximate (3.10b) by

$$(4.1) \quad \phi_i \frac{c_i^{n+1} - c_i^n}{\Delta t}$$

$$- \frac{\left[D(\tilde{u}_{i+1/2}^{n+1}) \dfrac{c_{i+1}^n - c_i^n}{\Delta x_{i+1/2}} - \tilde{u}_{i+1/2}^{n+1} c_{i+1/2}^n \right] - \left[D(\tilde{u}_{i-1/2}^{n+1}) \dfrac{c_i^n - c_{i-1}^n}{\Delta x_{i-1/2}} - \tilde{u}_{i-1/2}^{n+1} c_{i-1/2}^n \right]}{\Delta x_i}$$

$$= \tilde{c}_i^n q_i^n,$$

where (neglecting the gravity term in $\tilde{u}$)

$$(4.2) \quad \tilde{u}_{i+1/2}^{n+1} = -\frac{k_{i+1/2}}{\mu(c_{i+1/2}^n)} \frac{p_{i+1}^{n+1} - p_i^{n+1}}{\Delta x_{i+1/2}},$$

$$(4.3) \quad k_{i+1/2} = \left[\frac{1}{\Delta x_i + \Delta x_{i+1}} \left(\frac{\Delta x_i}{k_i} + \frac{\Delta x_{i+1}}{k_{i+1}} \right) \right]^{-1} \quad \text{(harmonic average)},$$

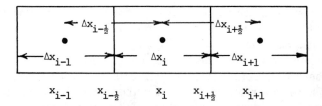

FIG. 4. *Block-centered finite difference coordinates.*

(4.4) $$c_{i+1/2}^n = \begin{cases} c_i^n, & \text{if } p_i^{n+1} > p_{i+1}^{n+1}, \\ c_{i+1}^n, & \text{if } p_i^{n+1} < p_{i+1}^{n+1} \end{cases} \quad \text{(upstream value)}.$$

We shall have more to say about block-centered differences and the velocity approximation (4.2) in §5, and about (4.4) (known in the industry as *upstream weighting*) in §6. For now, simply note that (4.1) can indeed be solved algebraically for c_i^{n+1}. If we multiply (4.1) by $\Delta x_i \Delta t$ and sum over i, then all spatial terms cancel (recall no-flow boundary conditions), and we obtain global conservation of mass:

(4.5) $$\sum_i \phi_i c_i^{n+1} \Delta x_i = \sum_i \phi_i c_i^n \Delta x_i + \sum_i \tilde{c}_i^n q_i^n \Delta t.$$

The key to an IMPES formulation is the absence of saturation (here, concentration) time derivatives in the pressure equation. More complicated systems [67], [4], [102] are handled by taking an appropriate linear combination of the component conservation equations to make these terms disappear. Computationally, IMPES is fast because only one variable (pressure) is solved for implicitly via a system of linear or nonlinear equations. However, the explicit scheme in (4.1) is subject to a Courant–Friedrichs–Lewy stability condition of the form $|\tilde{u}|\Delta t/\phi\Delta x \leq 1$ (possibly more stringent with significant nonlinearities), since the equation is almost hyperbolic. In the industry, this is known as a "throughput" condition, because it says that no more than one pore volume of material can be put through a grid block in one time step. In our work, we are simulating rather complex enhanced recovery techniques that require fairly fine grid resolution, particularly near wells and steep fronts. Thus, IMPES would restrict our time steps severely, and for this reason we chose not to use it. Another reason is that we are interested in finite element methods, and the accumulation integral analogous to the first term of (4.1) would have the form

$$\int_\Omega \phi c^{n+1} v \, dx$$

for test function v; this is nonzero for basis functions on adjacent elements, thus coupling them and leading to an implicit system of equations, even when all spatial terms are lagged at time level n. A finite element implementation of IMPES would have to lump the accumulation term such that it looked like its finite difference analogue.

The fully implicit procedure does not require a pressure equation and usually proceeds from a system of the form (3.9). Essentially, one has two equations like (4.1), except that n is replaced by $n + 1$ in the spatial and source terms. This nonlinear system is solved simultaneously for p (equivalently, $\tilde{u}$) and c by Newton's method. Each nonlinear iteration, typically 3 to 5 per time step, involves the solution of fully coupled linear equations for p and c. Mass conservation follows as in the IMPES case, up to the residual tolerance in the Newton iteration. For a system as simple as (3.9), it would actually be simpler to reduce to (3.10) for fully implicit as well as IMPES; in more complicated multiphase problems, with nonlinear relative permeability coefficients in the convection terms, separate pressures differing by saturation-dependent capillary pressures, and variable density and porosity in the accumulation terms, this is no longer the case.

Fully implicit time stepping eliminates the stability concerns of IMPES and makes accuracy the prime criterion setting time-step size, though one still has to take care that nonlinearities not prevent Newton's method from converging. For us, this procedure has two drawbacks. First, it is expensive—linear systems of order equal to the number of equations (two here, but greater in more complex problems) times the number of nodes must be solved several times at each time step. Because of this, virtually all chemical and compositional models in industry use the IMPES formulation. Second, when the equations are coupled and solved simultaneously, one is forced to use the same numerical method, grid, and time step for all equations. As we have seen, the equations in (3.10), if thought of as decoupled, exhibit quite different behavior, and one would like the flexibility of choosing separate and possibly different methods.

For these reasons, we decided to use sequential methods. A sequential method, like IMPES, uses a system like (3.10) and solves implicitly for p, but it also solves implicitly for c in a decoupled fashion. This means that either c in (3.10a) or $\tilde{u}$ in (3.10b) must be lagged or extrapolated in time to decouple the equations. Since the streamlines of a displacement should not change as rapidly as the movement of the front, we expect $\tilde{u}$ to be smoother in time than c; hence, we chose to extrapolate $\tilde{u}$ in (3.10b). Thus we solve first for c, then p (or $\tilde{u}$). In fact, this allows multiple concentration time steps for each pressure time step. Specific schemes will be discussed in subsequent sections, but a general picture has the form

$$(4.6a) \quad \phi \frac{c^{n+\beta} - c^{n+\alpha}}{(\beta - \alpha)\Delta t} - \nabla \cdot (D(E\tilde{u}^{n+\beta})\nabla c^{n+\beta} - (E\tilde{u}^{n+\beta})c^{n+\beta})$$

$$= \tilde{c}^{n+\beta} q^{n+\beta}, \quad 0 < \cdots < \alpha < \beta < \cdots < 1,$$

(4.6b)
$$-\nabla \cdot \left(\frac{k}{\mu(c^{n+1})}\nabla p^{n+1}\right) = q^{n+1},$$

where α and β are fractional indices ($\alpha = 0$ and $\beta = 1$ if both equations use the same time step) and $E\tilde{u}^{n+\beta}$ is a linear extrapolation from $\tilde{u}^{n-1}$ and $\tilde{u}^n$. Conservation of mass again follows. With this decoupling, both equations are linear; for more complex nonlinear systems, a Newton iteration is needed for both, and its residuals determine the mass-conservation error.

We note here that if $\alpha = 0$ and $\beta = 1$ in (4.6), we could iterate the procedure by successive substitution to reach the fully implicit solution; that is, replace $E\tilde{u}^{n+1}$ by the first computed $\tilde{u}^{n+1}$, solve for c^{n+1} and $\tilde{u}^{n+1}$ again, and so on. Thus, whatever stability and rates of convergence we can prove for our sequential procedures will apply also to the fully implicit method, provided that its Newton iteration can be shown to converge to a unique solution.

5. Methods for the pressure equation. Since we are using sequential time stepping, we can consider methods for (3.10a) and (3.10b) (or (3.14b)) separately. This section deals with procedures applied to the pressure equation (3.10a). As background, we begin with descriptions of some standard procedures. For economy of notation, we consider the problem

(5.1) $\qquad -\nabla \cdot (a(x)\nabla p) \equiv \nabla \cdot u = q, \qquad x \in \Omega,$

(5.2) $\qquad -(a\nabla p) \cdot \nu \equiv u \cdot \nu = 0, \qquad x \in \partial\Omega;$

lower-order gravity terms do not affect the methods or the mathematical analysis in any significant way, though they do influence physical flow patterns. For simplicity, let Ω be a unit square in $\mathbb{R}^2$. Take $a(x)$ (representing k/μ) to be possibly discontinuous, but bounded below by a positive constant. As noted in §3, compatibility demands that $\int_\Omega q\, dx = 0$, and p is only determined up to an additive constant. We can add the additional condition $\int_\Omega p\, dx = 0$ to fix p.

In the petroleum industry, finite differences are used almost universally. The version best known to mathematicians, called point-centered differences in the industry, consists of choosing meshes $0 = x_0 < x_1 < \cdots < x_k = 1$ and $0 = y_0 < y_1 < \cdots < y_l = 1$, defining interfaces $x_{i+1/2} = (x_i + x_{i+1})/2$ and $y_{j+1/2} = (y_j + y_{j+1})/2$, and approximating (5.1) at (x_i, y_j) by

(5.3) $\qquad \dfrac{-1}{x_{i+1/2} - x_{i-1/2}}\left[a_{i+1/2,j}\dfrac{p_{i+1,j} - p_{i,j}}{x_{i+1} - x_i} - a_{i-1/2,j}\dfrac{p_{i,j} - p_{i-1,j}}{x_i - x_{i-1}}\right]$

$\qquad -\text{(similar terms in } y\text{-direction)} = q_{i,j}, \qquad 0 \leq i \leq k, \ 0 \leq j \leq l.$

This couples (x_i, y_j) to its four immediate neighbors, and hence is called a five-point scheme. The boundary condition (5.2) may be handled by reflection, requiring that $p_{-1,j} = p_{1,j}$ for all j, and so on. The point-centered procedure is used in a small minority of industry simulators.

The vast majority of working codes use the block-centered method mentioned in §4. This idea is derived rather naturally from physical considerations as follows. Return to Fig. 4 in §4, and attach constant dimensions Δy and Δz to the blocks in the other two coordinates; for petroleum engineers, problems that are mathematically one-dimensional are still physically three-dimensional. Since $-a\nabla p$ represents a Darcy velocity, or volumetric flux per unit cross-sectional area per unit time, we can express the influx to block i from block $i+1$ in time Δt as

$$V_R = a_{i+1/2} \frac{p_{i+1} - p_i}{\Delta x_{i+1/2}} \Delta y \Delta z \Delta t.$$

The coefficient is regarded as a step function, constantly a_i and a_{i+1} on blocks i and $i+1$, respectively, and if one thinks of the time to flow from x_{i+1} to x_i as the sum of times from x_{i+1} to $x_{i+1/2}$ and $x_{i+1/2}$ to x_i, it makes sense to take $a_{i+1/2}$ as a harmonic average. Similarly, the influx to block i from block $i-1$ is

$$V_L = -a_{i-1/2} \frac{p_i - p_{i-1}}{\Delta x_{i-1/2}} \Delta y \Delta z \Delta t.$$

Finally, q_i is the volumetric injection per unit volume per unit time into block i, so in time Δt we have

$$V_I = q_i \Delta x_i \Delta y \Delta z \Delta t.$$

The steady-state equation (5.1) requires that

$$V_R + V_L + V_I = 0.$$

When extended to two-dimensional flow, this leads to a difference equation exactly like (5.3), except that the relationships between coordinates are different; in the point-centered case we have $x_{i+1/2} - x_{i-1/2} = (x_{i+1} - x_{i-1})/2$, but this does not necessarily hold in the block-centered case. Also, the block-centered version of (5.3) holds for $1 \leq i \leq k$, $1 \leq j \leq l$ with $x_{1/2} = 0$, $x_{k+1/2} = 1$, $y_{1/2} = 0$, $y_{l+1/2} = 1$ being boundary coordinates, and no-flow boundary conditions are easily incorporated by setting $a_{1/2,j} = a_{k+1/2,j} = a_{i,1/2} = a_{i,l+1/2} = 0$. This convenience at the boundary and the physical concept of flow between cells are the reasons why petroleum engineers prefer block-centered to point-centered schemes.

For uniform mesh, it is easy to see that the two procedures are identical except for their treatment of boundary conditions (point-centered puts nodes on the boundary, block-centered does not), and that their local truncation error is of order $(\Delta x)^2$. For smooth a and q, the analysis of Gerschgorin [38] yields $O((\Delta x)^2)$ error estimates at the nodes (x_i, y_j). The results are based on a maximum principle, and the exact solution is required to have four continuous derivatives. For Dirichlet boundary conditions with point-centered differences, stronger theorems needing at most two continuous derivatives have been demonstrated by Bramble, Hubbard, and Thomee for $q = 0$ [8], by Nitsche and Nitsche [62], by Kellogg [51], and by Bramble [7].

For nonuniform mesh, local truncation errors become interesting. Considering the simple equation $-p'' = q$, one can see for either approximation that, by Taylor expansions,

$$\frac{p_{i+1} - p_i}{x_{i+1} - x_i} - \frac{p_i - p_{i-1}}{x_i - x_{i-1}} = \frac{x_{i+1} - x_{i-1}}{2} p_i'' + O(\Delta x).$$

Both schemes divide by $x_{i+1/2} - x_{i-1/2}$ to approximate p_i''; in the point-centered case, this is equal to $(x_{i+1} - x_{i-1})/2$ and results in first-order correctness, but in the block-centered case equality does not hold (for example, if blocks $i - 1$ and $i + 1$ are longer than block i, $x_{i+1/2} - x_{i-1/2}$ will be less than $(x_{i+1} - x_{i-1})/2$) and the local approximation is inconsistent. This was first noted in the petroleum engineering literature by Settari and Aziz [86], [87], and there was concern about the convergence of block-centered differences, though they continued in common use. Settari and Aziz suspected that the procedure was convergent in spite of its inconsistency, and this has recently been proved in one dimension by Weiser and Wheeler [108] and by Kreiss et al. [53]. We shall show that the block-centered scheme is equivalent to a certain mixed finite element method (to be discussed later in this section) with a particular quadrature rule, and first-order convergence follows, even for discontinuous coefficient in multiple dimensions. Manteuffel and White [60] have second-order convergence theorems for various point-centered and block-centered schemes on nonuniform meshes, but this is not available in manuscript form at this writing, and we are not sure what schemes, dimensions, coefficient types, and right-hand sides have been covered. This extends early work of Samarskii [84] on second-order convergence for finite difference methods that were locally first-order.

Most of our work deals with finite element methods. Here we establish notation, describe standard Galerkin procedures, and recapitulate well known theorems that can be found in texts such as Ciarlet [14], Oden and Reddy [63], and Strang and Fix [97]. For simplicity we confine ourselves to tensor-product meshes of rectangles.

Choose meshes $\Delta_x: 0 = x_0 < x_1 < \cdots < x_k = 1$ and $\Delta_y: 0 = y_0 < y_1 < \cdots < y_l = 1$ as in the point-centered difference method. Define the piecewise-polynomial space

$$\mathcal{M}_q^r(\Delta) = \{v \in C^q([0,1]) : v \text{ is a polynomial}$$
$$\text{of degree} \leq r \text{ on each subinterval of } \Delta\},$$

where $q = -1$ refers to discontinuous functions. Thus, continuous piecewise bilinears on the square would be the tensor-product space $\mathcal{M}_0^1(\Delta_x) \otimes \mathcal{M}_0^1(\Delta_y)$. A basis for this space would be $\{v_i w_j : 0 \leq i \leq k, 0 \leq j \leq l\}$, where $v_i \in \mathcal{M}_0^1(\Delta_x)$ is 1 at x_i and 0 at all other nodes, and similarly for $w_j \in \mathcal{M}_0^1(\Delta_y)$. Define also the Sobolev space $H^m(\Omega) = \{v : \text{all partial derivatives of } v \text{ of order } \leq m \text{ are in } L^2(\Omega)\}$ and the norms

$$\|v\| = \|v\|_0 = \|v\|_{L^2(\Omega)} = \left(\int_\Omega |v|^2\, dx\, dy\right)^{1/2},$$

$$\|v\|_m = \|v\|_{H^m(\Omega)} = \left(\sum_{|\alpha| \leq m} \int_\Omega \left|\frac{\partial^{|\alpha|} v}{\partial x^\alpha}\right|^2\right)^{1/2}.$$

Let $(v, w) = \int_\Omega vw\, dx\, dy$ denote the usual L^2 inner product. In a Ritz variational method for (5.1)–(5.2), one notes that p is the unique function (of mean value zero) in $H^1(\Omega)$ minimizing the functional (representing energy in structural problems) $A(u) = \frac{1}{2}(a\nabla u, \nabla u) - (q, u)$, chooses a finite-dimensional subspace $\mathcal{M}$ of $H^1(\Omega)$, and finds the approximation P that minimizes A in $\mathcal{M}$. It is well known that this is equivalent to the Galerkin method:

Find $P \in \mathcal{M}$ (of mean value zero) such that

(5.4) $\qquad B(P, v) \equiv (a\nabla P, \nabla v) = (q, v), \quad \text{all } v \in \mathcal{M},$

where $(a\nabla P, \nabla v) = \int_\Omega a\nabla P \cdot \nabla v\, dx$. The Galerkin method is obtained by multiplying (5.1) by v and integrating by parts, using Green's theorem and (5.2). For nonselfadjoint equations, such as the convection-diffusion equation (3.10b), the Ritz variational minimization can no longer be defined, but the Galerkin procedure can be extended. In general, the space of trial functions (where the solution P is sought) may differ from the space of test functions v, though the dimensions must be equal. These spaces may relate to meshes of almost arbitrary character, allowing highly irregular geometries to be treated.

To compute with (5.4), choose a basis $\psi_1, \ldots, \psi_N$ for $\mathcal{M}$. Setting $v = \psi_I$, $I = 1, \ldots, N$, yields N equations for the N degrees of freedom in P. Thus, basis functions ψ_I and ψ_J are coupled by the matrix coefficient $(a\nabla \psi_I, \nabla \psi_J)$. For general a, this must be computed by a numerical integration rule or by projecting a into a finite-dimensional piecewise-polynomial space. Further, if we take $\mathcal{M} = \mathcal{M}_0^1(\Delta_x) \otimes \mathcal{M}_0^1(\Delta_y)$ for example, this coefficient is nonzero if ψ_I and ψ_J are supported on any common subrectangle; since each is supported on the four rectangles meeting the node where it is 1, a fixed ψ_I will be coupled to nine basis functions, including itself, its four immediate neighbors, and its four diagonal neighbors. In three dimensions, the coupling would extend to 27 points instead of the 7 with standard finite differences. Numerical integration and more extensive coupling make standard Galerkin procedures costlier than finite differences, a principal reason they have not made inroads in petroleum reservoir simulation. To justify their use, one must show that they are better equipped to handle certain problems. The usual finite element advantages of more flexible geometry and better treatment of boundary conditions mean little in reservoir simulation, since geometric irregularities rarely go beyond tilting of geological layers (easily handled by finite differences), and boundary effects are normally unimportant or easily modeled. One specific application that does suggest finite elements is simulation of hydraulic fracture propagation and subsequent flow; these fractures are created in low-permeability formations to enhance production [50], leading to questions of rock mechanics and irregular geometry. We shall not discuss this subject further here.

We see that we shall have to demonstrate some tangible benefits of finite element methods to offset their cost. First, we recall convergence rates for standard Galerkin procedures. In approximation theory, one shows that a piecewise polynomial P of degree $\leq r$ can approximate a smooth function p in $L^2(\Omega)$ and $H^1(\Omega)$ such that

$$\|p - P\| = O(h^{r+1}\|p\|_{r+1}), \qquad \|p - P\|_1 = O(h^r\|p\|_{r+1}),$$

where h is the largest linear dimension in the mesh underlying P. Then in Galerkin theory, one proves that the approximation selected by the Galerkin method comes within a constant multiple (independent of h and p) of minimizing the error. Putting these together, we see that for piecewise-bilinear approximations, for example, we can expect second-order convergence of pressures and first-order of Darcy velocities; for piecewise biquadratics, we would have third-order and second-order, respectively. The bilinear results could be matched by finite differences if the theorems of Bramble [7] could be extended to nonuniform mesh, Neumann boundary conditions, discontinuous coefficients, and so on. However, assuming $p \in H^3(\Omega)$, the higher-order convergence of biquadratics would have to make them more efficient than five-point finite differences on a sufficiently fine mesh. Even with pressure singularities at wells, local estimates away from wells could lead to similar conclusions. Thus, one can consider finite element methods with quadratic or higher-degree polynomials.

Young [116] does this in a clever way. On a tensor-product mesh, he uses Lagrange polynomials of degree r, $r = 1, 2$, or 3, as basis functions and integrates the matrix coefficients with an $(r + 1)$-point Lobatto quadrature rule (exact for polynomials of degree $\leq 2r - 1$). Since each Lagrange polynomial vanishes at all Lobatto points except one, the computational work of numerical integration is cut significantly. Yet every integral $(a\nabla\psi_I, \nabla\psi_J)$ would be computed exactly if a were constant, so by the theory of quadrature error in Galerkin methods [97], no orders of convergence are lost. For $r = 1$, Young's method is the same as point-centered finite differences (thus proving second-order convergence, even on nonuniform mesh), but he found $r = 2$ to be more efficient for the reservoir problems tested (miscible displacement among them), and $r = 3$ still more efficient. Hence it appears that Galerkin methods may indeed be competitive in reservoir simulation.

Our work relating to the pressure equation deals with other finite element concepts, *removal of singularities* and *mixed methods*. To describe removal of singularities for the pressure equation [19], [46], recall that the Laplace operator in two dimensions has the fundamental solution

$$p(x, y) = \frac{1}{2\pi} \log \sqrt{x^2 + y^2} = \frac{1}{2\pi} \log r$$

satisfying

(5.5) $$\Delta p = \delta_{(0,0)},$$

where $\delta_{(x,y)}$ denotes the Dirac measure with mass at (x, y). Thus, if a is constant and $q = Q\delta_{(0,0)}$, (5.1) has a solution

(5.6) $$p(x, y) = -\frac{Q}{2\pi a} \log r.$$

If we consider flow across the circle of radius R, (5.6) yields

$$-a\nabla p = \frac{Q}{2\pi R}$$

on the circle, which simply verifies that a total flux of Q distributed radially must lead to a Darcy velocity (flux per unit cross-section) of $Q/2\pi R$. In general, a is not constant, but its value at a well determines the leading singular behavior of the pressure around the well.

For miscible displacement, q will be of the form

(5.7) $$q = \sum_{i=1}^{w} Q_i \delta(x_i, y_i),$$

where w is the number of wells and well i is at (x_i, y_i). Let $r_i(x, y)$ be the distance from (x, y) to (x_i, y_i). Then the singular part of the pressure is

(5.8) $$p'(x, y) = -\frac{1}{2\pi} \sum_{i=1}^{w} \frac{Q_i}{a_i} \log r_i(x, y) \equiv \sum_{i=1}^{w} p'_i(x, y),$$

where $a_i = a(x_i, y_i)$. We seek a finite element method to obtain an approximation P'' of the regular part of the pressure,

(5.9) $$p''(x, y) = p - p'.$$

As we shall see theoretically and numerically, we can approximate p'' much better than we could approximate p.

A calculation using (5.1), (5.5), and (5.8) shows that

$$-\nabla \cdot (a(x)\nabla p'') = \nabla \cdot (a\nabla p') - \nabla \cdot (a\nabla p)$$
$$= \nabla a \cdot \nabla p' + a\Delta p' + q$$
$$= \nabla a \cdot \nabla p'$$
$$= \sum_i \nabla a \cdot \nabla p'_i$$

(5.10)
$$= \sum_i \nabla (a - a_i) \cdot \nabla p'_i$$
$$= \sum_i \nabla \cdot ((a - a_i)\nabla p'_i) - \sum_i (a - a_i)\Delta p'_i$$
$$= \sum_i \nabla \cdot ((a - a_i)\nabla p'_i), \qquad x \in \Omega;$$

for discontinuous a (permeability) this derivation, and the singular functions p'_i, should be confined to neighborhoods of the wells where a is continuous. Then p'' satisfies (5.10) and

(5.11) $\qquad -(a\nabla p'') \cdot v = (a\nabla p') \cdot v, \qquad x \in \partial\Omega.$

A variational form of (5.10)–(5.11) is, by Green's theorem,

(5.12)
$$(a\nabla p'', \nabla v) = -\sum_i ((a - a_i)\nabla p'_i, \nabla v)$$
$$- \sum_i \int_{\partial\Omega} (a_i\nabla p'_i \cdot v)v \, ds, \qquad v \in H^1(\Omega).$$

The right-hand side of (5.12) is well behaved, because $a - a_i$ should cancel the $1/r$ singularity of $\nabla p'_i$, and $\nabla p'_i \cdot v$ is zero near a boundary well if the boundary passes straight through. If Ω is a square, then $\nabla p'_i \cdot v$ is zero on an entire boundary edge containing well i. If also a is constant near wells, then the right-hand side of (5.10) is smooth, and p'' attains whatever degree of smoothness the domain Ω permits in the theory of elliptic boundary-value problems. For a square, $p'' \in H^2(\Omega)$, so second-order convergence of piecewise-bilinear finite elements (finding P'' in $\mathcal{M}$ satisfying (5.12) for all $v \in \mathcal{M}$, $\mathcal{M} = \mathcal{M}_0^1(\Delta_x) \otimes \mathcal{M}_0^1(\Delta_y)$) is established. If $a - a_i$ only vanishes to first order at well i, one can show that $p'' \in H^{2-\varepsilon}(\Omega)$ for any $\varepsilon > 0$, and that the above order of accuracy is attained if the next singular term of p is subtracted out. We shall not define fractional-index Sobolev spaces here; see, for example, Adams [1].

In contrast, finite differences cannot mimic this in a straightforward manner. The second derivative in the last member of (5.10) would have to be approximated by a difference quotient. If a is not constant, this may be singular at the well. The variational formulation enables one to remove the second derivative and handle any singularities via integrals. The usual finite difference technique is to distribute the source term q over a grid block, compute a grid-block pressure that can be associated with an effective radius [68], [69], and use a well model such as (3.19) of Chapter I to obtain a pressure surface near the well. We have not attempted to analyze this theoretically, but it seems clear that it should be less accurate than the finite element method with singularities removed.

The idea of mixed methods for the pressure equation is to approximate the pressure and velocity simultaneously in a variational method, rather than to obtain velocity by differencing (as in (4.2)) or differentiating (using $-a\nabla(p' + P'')$) pressure. It is an elementary consequence of approximation theory that one order of convergence is lost in passing from pressure to velocity by differentiation. Additionally, the coefficient a may be rough (because of variations in viscosity, depending on concentration) or even discontinuous (because of permeability), but velocity is smooth. In the system (3.10), pressure is unimportant in the concentration equation; velocity is what is needed. In a compressible system, pressure would also appear, and in more complicated systems, viscosity and interphase mass transfer would also be functions of pressure. Because of these considerations, it would be desirable to have procedures that approximate both pressure and velocity to the same order of accuracy. This is what mixed methods do.

We begin by separating (5.1) into two equations representing the two physical principles that led to it:

(5.13) $\qquad u = -a\nabla p, \qquad x \in \Omega \quad$ (Darcy's law),

(5.14) $\qquad \nabla \cdot u = q, \qquad x \in \Omega \quad$ (conservation of mass).

Let $H(\text{div}; \Omega)$ be the set of vector functions $v \in (L^2(\Omega))^2$ such that $\nabla \cdot v \in L^2(\Omega)$ and let

(5.15) $\qquad V = \{v \in H(\text{div}; \Omega) : v \cdot \nu = 0 \text{ on } \partial\Omega\}.$

Let $W = L^2(\Omega)$, and denote the vector inner product on $(L^2)^2$,

$$(v, w) = \int_\Omega v \cdot w \, dx,$$

in the same way as the scalar inner product on L^2. To obtain a variational form of (5.13)–(5.14), multiply (5.13) by $v \in V$, divide by a, integrate, and integrate by parts to see that

(5.16) $\qquad (a^{-1}u, v) - (p, \nabla \cdot v) = 0, \qquad v \in V.$

Next, multiply (5.14) by $w \in W$ and integrate to obtain

(5.17) $\qquad (\nabla \cdot u, w) = (q, w), \qquad w \in W.$

The system (5.16)–(5.17) is then approximated by finite elements. For suitable finite-dimensional subspaces $V_h \subset V$ and $W_h \subset W$ to be described next, the approximation is $\{U, P\} \in V_h \times W_h$ satisfying

(5.18) $\qquad (a^{-1}U, v) - (P, \nabla \cdot v) = 0, \qquad v \in V_h,$

(5.19) $\qquad (\nabla \cdot U, w) = (q, w), \qquad w \in W_h.$

The finite-dimensional subspaces need certain properties in order for the convergence analysis of Brezzi [9], Falk and Osborn [35], and Raviart and Thomas [79] to hold. One of these is that div $V_h \subset W_h$, which makes the somewhat strange nature of the following spaces, developed by Raviart and Thomas, understandable. On a rectangular mesh over the unit square described by Δ_x and Δ_y as before, let the spaces of index r, $r = 0, 1, 2, \cdots$, be

$$W_h^r = \mathcal{M}_{-1}^r(\Delta_x) \otimes \mathcal{M}_{-1}^r(\Delta_y),$$
$$\tilde{V}_h^r = [\mathcal{M}_0^{r+1}(\Delta_x) \otimes \mathcal{M}_{-1}^r(\Delta_y)] \times [\mathcal{M}_{-1}^r(\Delta_x) \otimes \mathcal{M}_0^{r+1}(\Delta_y)],$$

(5.20) $\qquad V_h^r = \{v = (v_x, v_y) \in \tilde{V}_h^r : v_x(0, y)$
$$= v_x(1, y) = 0, v_y(x, 0) = v_y(x, 1) = 0\}$$
$$= \{v \in \tilde{V}_h^r : v \cdot \nu = 0 \text{ on } \partial\Omega\},$$

where h measures the largest linear dimension in the mesh. Thus, the pressure approximation is discontinuous at mesh edges, and the x- (y-) component of the velocity approximation is discontinuous at edges parallel to the x- (y-) axis. Raviart and Thomas also developed spaces for use on triangular meshes, which we shall not discuss. For smooth right-hand side q, the analysis of [9], [35], [79] yields the error estimates

(5.21) $\qquad \|p - P\|_{L^2} = O(h^{r+1}),$
$\qquad\qquad \|u - U\|_V = O(h^{r+1}),$

where the norm in V is defined by

$$\|v\|_V = \|(v_x, v_y)\|_V = (\|v_x\|^2 + \|v_y\|^2 + \|\nabla \cdot v\|^2)^{1/2}.$$

As desired, the errors in pressure and velocity are of the same order.

For singular q, we can remove logarithms from the pressure as in the Galerkin case. However, it may be of more interest to remove terms of order $Q/2\pi r$ from the velocity and obtain more accurate velocity approximations. We describe this idea, first suggested by D. N. Arnold, here. Represent the velocity vector u as the sum of a singular part u' and a regular part u'', with

(5.22) $\qquad u' = \dfrac{1}{2\pi} \sum_{i=1}^{w} Q_i \nabla (\log r_i),$

using the notation from the Galerkin case. Then $\nabla \cdot u' = q$, so that u'' satisfies

(5.23) $$\nabla \cdot u'' = 0, \quad x \in \Omega,$$

(5.24) $$u'' \cdot \nu = -u' \cdot \nu, \quad x \in \partial\Omega.$$

We also have

(5.25) $$u'' = -a\nabla p - u', \quad x \in \Omega,$$

and (5.23) through (5.25) replace the original (5.2), (5.13), and (5.14). As before, multiply (5.25) by $v \in V$, divide by a, integrate, and integrate by parts, yielding

(5.26) $$(a^{-1}u'', v) - (p, \nabla \cdot v) = -(a^{-1}u', v), \quad v \in V,$$

and multiply (5.23) by $w \in W$ and integrate to obtain

(5.27) $$(\nabla \cdot u'', w) = 0, \quad w \in W.$$

Then (5.24), (5.26), and (5.27) comprise a variational form that can be discretized with the Raviart–Thomas spaces V_h^r, W_h^r. This is formulated in analogy with (5.18) and (5.19), except that U'' is no longer sought in the space V_h^r (normal component zero), but rather should lie in $\tilde{V}_h^r$ and have normal component as close as possible to that of $-u'$. This is accomplished by enforcing the condition

(5.28) $$\int_{\partial\Omega} (u' + U'') \cdot \nu(v \cdot \nu)\, ds = 0, \quad v \in \tilde{V}_h^r.$$

If $r = 0$, then $v \cdot \nu$ is constant on each boundary interval, so this says that $U'' \cdot \nu$ must take the average value of $-u' \cdot \nu$ on each interval. We remark that U'' can be defined alternatively on $\partial\Omega$ by interpolating $-u'$ at appropriate Gauss points on $\partial\Omega$.

For this procedure with singular q, Douglas, Ewing, and Wheeler have extended the analysis of [9], [35], [79] for consistency with earlier syntax to prove that

(5.29) $$\|p - P\| = O(h \log 1/h),$$

(5.30) $$\|u - (u' + U'')\| = O(h \log 1/h).$$

Numerical results presented in [34] for $r = 1$ showed first-order convergence for p and u on Ω and second-order convergence away from singularities, as theory would predict. Without removal of singularities, velocity failed to converge on Ω (otherwise, results were the same). This is expected, because $p \in H^{1-\varepsilon}(\Omega)$ and $u \in H^{0-\varepsilon}(\Omega)$ for any $\varepsilon > 0$.

At this stage it is appropriate to examine the linear matrix problem associated with the approximation of (5.26)–(5.27), since it is of a different form from the usual finite difference or Galerkin matrix. For simplicity we consider the case

$r = 0$. With the partitions Δ_x and Δ_y from before, let bases for subspaces be

$$\mathcal{M}_0^1(\Delta_x) : \{v_i^x : 1 \leq i \leq k - 1\}, \quad v_i^x(x_m) = \delta_{im},$$
$$\mathcal{M}_{-1}^0(\Delta_x) : \{w_i^x : 1 \leq i \leq k\}, \quad w_i^x(x) = 1 \text{ if } x_{i-1} < x < x_i,$$
$$\mathcal{M}_0^1(\Delta_y) : \{v_j^y : 1 \leq j \leq l - 1\}, \quad v_j^y(y_m) = \delta_{jm},$$
$$\mathcal{M}_{-1}^0(\Delta_y) : \{w_j^y : 1 \leq j \leq l\}, \quad w_j^y(y) = 1 \text{ if } y_{j-1} < y < y_j.$$

Then bases for $(V_h)_x$, $(V_h)_y$, and W_h are, respectively, $\{v_i^x w_j^y\}$, $\{w_i^x v_j^y\}$, and $\{w_i^x w_j^y\}$. Number the basis for $(V_h)_x$ such that i changes rapidly, and for $(V_h)_y$ such that j changes rapidly. A basis for V_h is $((V_h)_x \times 0) \cup (0 \times (V_h)_y)$. The resulting matrix problem has the form

$$(5.31) \qquad \begin{pmatrix} M_x & 0 & -N_x \\ 0 & M_y & -N_y \\ N_x^T & N_y^T & 0 \end{pmatrix} \begin{pmatrix} U_x'' \\ U_y'' \\ P \end{pmatrix} = \begin{pmatrix} R_x \\ R_y \\ RR \end{pmatrix},$$

where

$$(M_x)_{i'j',ij} = (a^{-1} v_i^x w_j^y, v_{i'}^x w_{j'}^y) \quad \text{(zero unless } j = j', |i - i'| \leq 1\text{),}$$
$$(M_y)_{i'j',ij} = (a^{-1} w_i^x v_j^y, w_{i'}^x v_{j'}^y) \quad \text{(zero unless } i = i', |j - j'| \leq 1\text{),}$$
$$(N_x)_{i'j',ij} = \left(w_{i'}^x w_{j'}^y, \frac{\partial}{\partial x}(v_i^x w_j^y)\right) \quad \text{(zero unless } j = j', i' = i \text{ or } i - 1\text{),}$$
$$(N_y)_{i'j',ij} = \left(w_{i'}^x w_{j'}^y, \frac{\partial}{\partial y}(w_i^x v_j^y)\right) \quad \text{(zero unless } i = i', j' = j \text{ or } j - 1\text{),}$$

and R_x, R_y, and RR contain the right-hand side of (5.26) (in R_x and R_y) together with boundary integrals from (5.28). The numbering system makes M_x and M_y tridiagonal, N_x bidiagonal. By permuting rows and corresponding columns N_y can be made bidiagonal. Since M_x and M_y are positive-definite and easily decomposed into lower- and upper-triangular factors, we can eliminate N_x^T and N_y^T in (5.31) and see that

$$(5.32) \qquad AP \equiv (N_x^T M_x^{-1} N_x + N_y^T M_y^{-1} N_y) P$$
$$= RR - N_x^T M_x^{-1} R_x - N_y^T M_y^{-1} R_y.$$

We solve (5.32), a system of order kl (essentially the same order as for a continuous piecewise-linear Galerkin method), for P and then back-substitute for U_x'' and U_y''. The matrix A is symmetric and positive-semidefinite, and we have used preconditioned conjugate gradient methods to solve the system. The choice of a good preconditioner is difficult and is a subject of current research. For higher-order spaces there are more diagonals in the matrix, but the basic ideas are the same. The nodes for the discontinuous bases are placed at Gauss quadrature points and appropriate Gauss rules are used to evaluate the integrals for M_x and M_y, minimizing the work of generating the matrix.

In this context it is worthwhile to note an analogy between the mixed method for $r = 0$ and block-centered finite differences. In both procedures, one obtains a pressure best regarded as constant on each grid block with a node at the center of the block. The finite difference method satisfies conservation of mass on each block, and the mixed-method velocity approximation satisfies the conservation equation $\nabla \cdot U = q$ everywhere. To see the latter statement, note that $\nabla \cdot U''$ is constant on each block, and taking w in (5.27) to be 1 on a given block and zero elsewhere, it follows that $\nabla \cdot U'' = 0$; since div $V_h^r \subset W_h^r$ (V_h^r and W_h^r defined in (5.20)), (5.27) with $w = \nabla \cdot U''$ implies that on each cell $\nabla \cdot U'' = 0$ pointwise.

To strengthen the analogy, we now wish to show that the block-centered finite difference method ((5.3) with the relationships of Fig. 4 in §4) is the lowest-order mixed method ((5.18)–(5.19) with $V_h = V_h^0$, $W_h = W_h^0$) with special numerical quadrature formulas. As in Fig. 4, partition [0, 1] with $\Delta_x : 0 = x_{1/2} < x_{3/2} < \cdots < x_{k+1/2} = 1$ and $\Delta_y : 0 = y_{1/2} < y_{3/2} < \cdots < y_{l+1/2} = 1$ and set $x_i = (x_{i-1/2} + x_{i+1/2})/2$, $y_j = (y_{j-1/2} + y_{j+1/2})/2$, $\Delta x_i = x_{i+1/2} - x_{i-1/2}$, $\Delta x_{i+1/2} = x_{i+1} - x_i$, $\Delta y_j = y_{j+1/2} - y_{j-1/2}$, $\Delta y_{j+1/2} = y_{j+1} - y_j$. Multiply (5.3) by $\Delta x_i \Delta y_j$, the area of a cell, to obtain

$$(5.33) \quad \Delta y_j((U_x)_{i+1/2,j} - (U_x)_{i-1/2,j}) + \Delta x_i((U_y)_{i,j+1/2} - (U_y)_{i,j-1/2}) = \Delta x_i \Delta y_j q_{ij},$$

where U_x and U_y are velocity components defined by

$$(5.34a) \quad (U_x)_{i+1/2,j} = -a_{i+1/2,j} \frac{P_{i+1,j} - P_{i,j}}{\Delta x_{i+1/2}},$$

$$(5.34b) \quad (U_y)_{i,j+1/2} = -a_{i,j+1/2} \frac{P_{i,j+1} - P_{i,j}}{\Delta y_{j+1/2}}.$$

Let U be the unique function in V_h^0 satisfying (5.34), and let P be the piecewise-constant function with cell values $P_{i,j}$. We show that $\{U, P\} \in V_h^0 \times W_h^0$ satisfies (5.18)–(5.19) if the integrals are evaluated in a certain way.

Note that the left-hand side of (5.33) is equal to

$$\Delta y_j \Delta x_i \frac{\partial}{\partial x}(U_x) + \Delta x_i \Delta y_j \frac{\partial}{\partial y}(U_y) = \int_{x_{i-1/2}}^{x_{i+1/2}} \int_{y_{j-1/2}}^{y_{j+1/2}} \nabla \cdot U,$$

and the right-hand side is the midpoint-rule integral of q over the cell. Thus, U satisfies

$$(5.35) \quad (\nabla \cdot U, w) = (q, w)_{M_x M_y}, \quad w \in W_h^0,$$

where $M_x M_y$ denotes midpoint-rule quadrature in both directions. Next, (5.34a) implies that

$$\frac{1}{2}(\Delta x_i + \Delta x_{i+1})\Delta y_j \frac{1}{a_{i+1/2,j}} (U_x)_{i+1/2,j} - \Delta y_j(P_{i,j} - P_{i+1,j}) = 0,$$

which can be rewritten as

(5.36) $\qquad (a^{-1}U_x, v^x_{i+1/2}w^y_j)_{T_xM_y} - \left(P, \dfrac{\partial}{\partial x}(v^x_{i+1/2}w^y_j)\right) = 0,$

where $v^x_{i+1/2}$ and w^y_j correspond to v^x_i (linear basis function) and w^y_j (constant basis function) in the discussion of the mixed method matrix, and T_xM_y denotes trapezoidal integration in the x-direction tensored with the midpoint rule in the y-direction. Similarly,

(5.37) $\qquad (a^{-1}U_y, w^x_i v^y_{j+1/2})_{M_xT_y} - \left(P, \dfrac{\partial}{\partial y}(w^x_i v^y_{j+1/2})\right) = 0.$

Combining (5.36) and (5.37), we obtain

(5.38) $\qquad (a^{-1}U_x, v_x)_{T_xM_y} + (a^{-1}U_y, v_y)_{M_xT_y} - (P, \nabla \cdot v) = 0, \qquad v \in V^0_h.$

By (5.35) and (5.38), the block-centered pressure and associated velocity satisfy the mixed-method equations (5.18) and (5.19), provided that the quadrature rules M_xM_y, T_xM_y, and M_xT_y are used as indicated.

The boundary conditions also agree. Because of the artifice $a_{1/2,j} = a_{k+1/2,j} = a_{i,1/2} = a_{i,l+1/2} = 0$ in the block-centered method, (5.34a) and (5.34b) yield $U \cdot v = 0$, which is the condition imposed by the space V^0_h in the mixed method. The relationship between block-centered differences and the mixed method with $r = 0$ is now demonstrated.

One can use (5.35), (5.38), error estimates for quadrature rules, and the analysis of mixed methods to obtain (5.21) with $r = 0$, i.e., if $\{U, P\}$ is obtained by block-centered differences and (5.34), then

(5.39) $\qquad \|p - P\| + \|u - U\|_V = O(h).$

This confirms mathematically the appropriateness of harmonically averaging the coefficient a, which was motivated physically in the derivation of block-centered differences from cell balances.

As noted in the discussion of Young's method with $r = 1$ [116], a similar analogy exists between point-centered differences and the bilinear Galerkin procedure with quadrature rule T_xT_y.

6. Methods for the concentration equation. This section considers approximation schemes for the concentration equation (3.10b) or (3.14b). We begin by cataloguing the standard methods used in the petroleum industry, together with some alternatives that have been discussed in the literature. Then we introduce the procedures that are the subject of our work.

Since we deal here with the concentration equation alone, assume that a Darcy velocity u is known, so that (3.10b), (3.11b), and (3.13) are

$$(6.1) \qquad \phi \frac{\partial c}{\partial t} - \nabla \cdot (D(u)\nabla c - uc) = \tilde{c}q, \qquad x \in \Omega, \quad t \in J,$$

$$(6.2) \qquad (D\nabla c) \cdot v - c(u \cdot v) = 0, \qquad x \in \partial\Omega, \quad t \in J.$$

$$(6.3) \qquad c(x, 0) = c_0(x), \qquad x \in \Omega.$$

The usual industry approaches to the system (6.1)–(6.3) entail block-centered finite differences with upstream weighting (see (4.1) through (4.4)), leaving out the diffusion-dispersion term (because, as we shall see, upstream weighting gives rise to numerical dispersion that exceeds physical dispersion on practical meshes). Time stepping may be either implicit or explicit; because our methods are implicit, we write the implicit scheme:

$$(6.4) \qquad \phi_{i,j} \frac{c_{i,j}^{n+1} - c_{i,j}^n}{\Delta t} + \frac{u_{i+1/2,j} c_{i+1/2,j}^{n+1} - u_{i-1/2,j} c_{i-1/2,j}^{n+1}}{\Delta x_i}$$
$$+ \frac{u_{i,j+1/2} c_{i,j+1/2}^{n+1} - u_{i,j-1/2} c_{i,j-1/2}^{n+1}}{\Delta y_j} = \tilde{c}_{i,j}^{n+1} q_{i,j}^{n+1},$$

where

$$(6.5) \qquad c_{i+1/2,j}^{n+1} = \begin{cases} c_{i,j}^{n+1} & \text{if } (u_{i+1/2,j})_x > 0, \\ c_{i+1,j}^{n+1} & \text{if } (u_{i+1/2,j})_x < 0, \end{cases}$$

and similarly in the y-direction for $c_{i,j+1/2}^{n+1}$. If $u = (u_x, u_y)$ and ϕ are constant, then as noted in (4.4) of Chapter I, the upstream weighting (6.5) corresponds to a numerical dispersion term of

$$(6.6) \qquad \nabla \cdot (D_{\text{num}} \nabla c) = \frac{|u_x| \Delta x}{\phi} \frac{\partial^2 c}{2 \partial x^2} + \frac{|u_y| \Delta y}{\phi} \frac{\partial^2 c}{2 \partial y^2}.$$

Upstream weighting is used in the industry because it suppresses nonphysical oscillations in the finite difference solution. As is well known, central differences and standard Galerkin procedures exhibit oscillations (also known as *overshoot*) when applied to convection-dominated flows (flows of high Peclet number—see the discussion in §3), especially at the trailing end of moving fronts. For several views on this topic, see the collection of papers in [49]. Petroleum engineers working in the field do not like to see oscillatory solutions, preferring any smooth solution to even a fairly accurate one with small wiggles. The standard response of reservoir model developers to this is to live with numerical dispersion, perhaps attempting to use grid spacing such that the coefficients in (6.6) correspond roughly to physical dispersion.

This approach is fraught with dangers. To understand the situation better, we shall attempt to represent (6.6) in terms of physical dispersion models. Suppose $\Delta x = \Delta y = h$, and let $v = u/\phi$ be the interstitial velocity. Then (6.6) can be written in tensor form as

$$(6.7) \qquad D_{num} = \frac{h}{2} \begin{pmatrix} |v_x| & 0 \\ 0 & |v_y| \end{pmatrix}.$$

If θ is the angle between v and the x-axis, we see from (3.8) that a physical dispersion model (neglecting diffusion because D_{num} is obviously velocity-dependent) has the form

$$(6.8) \qquad D_{phys} = \alpha_l |v| \begin{pmatrix} \cos^2\theta & \cos\theta\sin\theta \\ \cos\theta\sin\theta & \sin^2\theta \end{pmatrix}$$
$$+ \alpha_t |v| \begin{pmatrix} \sin^2\theta & -\cos\theta\sin\theta \\ -\cos\theta\sin\theta & \cos^2\theta \end{pmatrix}.$$

For different values of θ, we seek α_l and α_t such that $D_{num} = D_{phys}$. For general θ this is not possible because, even with symmetry, there are three equations in two unknowns. However, we do obtain [107]

$$(6.9) \qquad \begin{aligned} \theta &= 0, \pm\frac{\pi}{2}, \cdots : \alpha_l = \frac{h}{2}, \; \alpha_t = 0, \\ \theta &= \pm\frac{\pi}{4}, \pm\frac{3\pi}{4}, \cdots : \alpha_l = \alpha_t = \frac{h}{2\sqrt{2}} \approx 0.354\, h. \end{aligned}$$

We see considerable differences, depending on whether the velocity is parallel ($\theta = 0$) or diagonal ($\theta = \pi/4$) to the grid. Consider Figs. 3 and 4 of Chapter I. In the diagonal case, significant transverse dispersion ($h/2\sqrt{2}$) impedes the flow from the injector to the producer; in the parallel case, no such dispersion exists. This, in addition to any preferential flow along grid paths in a five-point difference scheme, makes the discrepancies in Figs. 4 and 5 of Chapter I understandable. Similar anomalies have been observed in thermal recovery by steam displacement [16].

We note also the implications for problems with high Peclet numbers, which may be in the tens or hundreds (see §2). If the mesh is uniform and the domain length is L, then $N = L/h$ is the number of grid intervals in one direction. Using $\alpha_l = h/2$, the Peclet number with numerical dispersion is

$$(6.10) \qquad Pe_{num} = L/(\alpha_l)_{num} = 2L/h = 2N.$$

Thus, accurate simulation of a problem with Peclet number in the hundreds, in three dimensions with several layers, could require 100,000 grid blocks. Adaptive refinement could reduce this greatly, but it seems more fruitful to look for better methods first.

One alternative suggested in [101] is two-point upstream weighting, in which (6.5) is replaced by (for uniform mesh)

$$(6.11) \qquad c_{i+1/2} = \begin{cases} \tfrac{3}{2}c_i - \tfrac{1}{2}c_{i-1}, & u_{i+1/2} > 0, \\ \tfrac{3}{2}c_{i+1} - \tfrac{1}{2}c_{i+2}, & u_{i+1/2} < 0. \end{cases}$$

The analysis leading to (6.6) would make this appear to be a second-order approximation, but in the presence of a sharp front it could be expected to be little better than single-point upstream weighting. Two-point methods have been helpful for certain problems, but as shown in [114] (see Fig. 5 of Chapter I), not for adverse-mobility-ratio displacements.

Yanosik and McCracken [114] simulated such problems with a nine-point finite difference scheme in two dimensions, using upstream weighting. Their method is based on the well-known fourth-order approximation of the Laplace operator on a uniform grid,

$$(6.12) \qquad (\Delta p)_{ij} \approx \frac{1}{6h^2} [-20 p_{ij} + 4(p_{i-1,j} + p_{i+1,j} + p_{i,j-1} + p_{i,j+1}) \\ + 1(p_{i-1,j-1} + p_{i-1,j+1} + p_{i+1,j-1} + p_{i+1,j+1})],$$

which one obtains by approximating $\tfrac{2}{3}$ of the Laplacian by the usual five-point operator and the other $\tfrac{1}{3}$ by the same operator rotated by 45° on a scale of $\sqrt{2}h$. They did essentially the same thing with the convection term $\nabla \cdot (uc) = -\nabla \cdot (c(k/\mu)\nabla p)$ of (6.1), using upstream weighting for the coefficient c multiplying each of the eight differences between p_{ij} and its neighbors. Thus, their numerical dispersion is $\tfrac{2}{3}$ of D_{num} from (6.7), plus a rotated term; it turns out to be

(6.13)
$$D_{\text{num}} = \frac{h}{12} \begin{pmatrix} 4|v_x| + |v_x + v_y| + |v_x - v_y| & |v_x + v_y| - |v_x - v_y| \\ |v_x + v_y| - |v_x - v_y| & 4|v_y| + |v_x + v_y| + |v_x - v_y| \end{pmatrix}.$$

They modified the scheme suitably for nonuniform grids. For uniform grids, we can do the same decomposition into physical dispersion terms as before, and we find

$$(6.14) \qquad \theta = 0, \pm\frac{\pi}{2}, \ldots : \alpha_l = \frac{h}{2}, \alpha_t = \frac{h}{6},$$

$$\theta = \pm\frac{\pi}{4}, \pm\frac{3\pi}{4}, \ldots : \alpha_l = \frac{\sqrt{2}}{3} h \approx 0.471h, \alpha_t = \frac{\sqrt{2}}{6} h \approx 0.236h.$$

The difference between $\theta = 0$ and $\theta = \pi/4$ is about one fifth of what it was for five-point differences. Reduced dependence on grid orientation should be expected, and this was observed. Numerical dispersion, however, is as severe as before.

Watts and Silliman [107] experimented with a refinement of each rectangular grid block into four triangles, by introducing the two diagonals into each rectangle. They used midpoint weighting (central differences) and a physical dispersion model with $\alpha_l = \alpha_t$. The triangles allow flow in diagonal directions, and the dispersion is not grid-oriented. Also, the triangles can be numbered such that the nonzero pattern of the system of linear equations is a subset of that for standard five-point differences; this makes the procedure attractive for implementation in existing simulators. For mobility ratio 41, rather coarse mesh (10 × 10), and fairly high physical dispersion (the units are not clear from the paper), grid orientation was very mild. It would be interesting to observe the behavior for lower dispersion, and particularly for the physically realistic case of $\alpha_l \gg \alpha_t$. If the method could give sharp, nonoscillatory fronts in such cases, it would be very attractive.

We consider next what can be done with fairly standard finite element methods in this area. As in §5, we can multiply (6.1) by a test function $v \in H^1(\Omega)$, then use Green's theorem and (6.2) to obtain the variational form

(6.15)
$$\left(\phi \frac{\partial c}{\partial t}, v\right) + (D\nabla c - uc, \nabla v) = (\tilde{c}q, v), \quad t \in J,$$
$$c(x, 0) = c_0(x), \quad x \in \Omega.$$

Then we choose a finite-dimensional subspace $\mathcal{M}$ of $H^1(\Omega)$, approximate c_0 by $C^0 \in \mathcal{M}$ (by, for example, interpolation or projection in the L^2 inner product—the latter conserves mass), and replace the time derivative $(\partial c/\partial t)^{n+1}$ by (say) a backward difference quotient over time step Δt to get an implicit time-stepping scheme: Find $C^1, C^2, \ldots \in \mathcal{M}$ such that

(6.16)
$$\left(\phi \frac{C^{n+1} - C^n}{\Delta t}, v\right) + (D\nabla C^{n+1} - uC^{n+1}, \nabla v) = (\tilde{C}^{n+1}q^{n+1}, v), \quad \text{all } v \in \mathcal{M}.$$

It is instructive to compare the accumulation terms of (6.16) and the finite difference method (6.4). In (6.16), neighboring nodes are coupled by integration over common elements, while they are not coupled in (6.4). On a uniform mesh in two dimensions, it can be shown that the piecewise-bilinear finite element matrix is what would be obtained with finite differences if the additional dispersion term

$$-\frac{h^2}{6}\frac{\partial}{\partial t}(\Delta c) + \frac{h^4}{36}\frac{\partial}{\partial t}\left(\frac{\partial^4 c}{\partial x^2 \partial y^2}\right)$$

were added to the left-hand side of (6.1). Thus, finite elements are slightly more dispersive than finite differences.

The theory of Galerkin methods for parabolic equations shows that, for smooth coefficients (including q) and domain (hence smooth solution c), the error $c - C$ is optimal in space (order h^{r+1} in L^2 for polynomials of degree $\leq r$) and time (first-order). To be more specific, for $w(x, t)$ defined on $\Omega \times J$, we define the norms

$$\|w\|_{L^2(J;H^m(\Omega))} = \left(\int_0^T \|w(\cdot, t)\|_{H^m(\Omega)}^2 dt\right)^{1/2},$$

$$\|w\|_{L^\infty(J;H^m(\Omega))} = \underset{0 \leq t \leq T}{\text{ess sup}} \|w(\cdot, t)\|_{H^m(\Omega)}.$$

Then the error satisfies [22, 109]

(6.17)
$$\max_n \|c^n - C^n\| \leq Kh^{r+1}\left(\|u\|_{L^\infty(J;H^m(\Omega))} + \left\|\frac{\partial u}{\partial t}\right\|_{L^2(J;H^{m-1}(\Omega))}\right)$$
$$+ K\Delta t \left\|\frac{\partial^2 u}{\partial t^2}\right\|_{L^2(J;L^2(\Omega))},$$

where K is independent of h, Δt, and u.

A few comments about this error estimate are in order. If (6.1) is strongly convection-dominated, a heuristic argument of the type leading to (6.6) and using $\phi\, \partial/\partial t \approx u\, \partial/\partial x$ shows that the one-sided time difference gives rise to a numerical dispersion term with coefficients $\alpha_l = |v|\Delta t/2$, $\alpha_t = 0$, where $v = u/\phi$. This corresponds to a physical model and hence is not grid-oriented, but it does require small time steps to avoid excessive numerical dispersion. (Note that an explicit method would have the same dispersion with negative sign, hence would be subject to a stability constraint on the time-step size.) Further, the nonsymmetric convection term causes difficulties in the error analysis, as terms arising from it have to be bounded by a combination of diffusive-dispersive terms with small coefficients and accumulation terms with large coefficients; these accumulation terms are ultimately eliminated by Gronwall's lemma, but the constant in (6.17) acquires exponential growth in time. Thus (6.17), while true, is not entirely satisfactory for convection-dominated flow. The oscillations and/or numerical dispersion characteristic of simulation of such problems are testimony to that.

Young [116] applied his Lagrange–Lobatto–Galerkin procedure (see §5) to the concentration equation as well as to the pressure equation, using a physical dispersion model with $\alpha_l = \alpha_t$. Because of the accumulation integral $(1/\Delta t)(\phi C^{n+1}, v)$ in (6.16), which would not be computed exactly with constant coefficients, one order of accuracy in space could be lost. This is likely more than offset at practical levels of discretization by the efficiencies of the quadrature, described in §5. The accumulation integral is nonzero only if the same basis function is substituted for C^{n+1} and v, so lumping is achieved and IMPES time stepping could be considered. For all polynomial degrees tested, including bilinears which are the same as point-centered five-point finite differences, Young found that oil recoveries from diagonal and parallel grids (as in, for example, Fig. 5 of Chapter I) converged to the same answer as the grids were refined. This is in contrast to the findings of [16] with upstream weighting and shows that grid-orientation sensitivity is due in large measure to the grid-dependent dispersion of upstream weighting, not merely to the lack of diagonal flow channels in five-point differences. Better results were found with biquadratics and bicubics. Small time steps were used, as dictated by Peclet numbers from about 50 to 300, and grids were fine enough to resolve fronts; presumably coarser grids would lead to oscillations.

Our work with the concentration equation involves *interior penalties* with finite elements, a *modified method of characteristics* with finite differences and finite elements, and a cell-balance finite element scheme that can be likened to the Yanosik–McCracken nine-point difference procedure.

The interior penalty Galerkin procedure was motivated by early computational work of Douglas and Dupont, described in [19], with continuous biquadratic polynomials ($\mathcal{M}_0^2(\Delta_x) \times \mathcal{M}_0^2(\Delta_y)$) and of Settari et al. [88] with continuously differentiable bicubic ($\mathcal{M}_1^3(\Delta_x) \times \mathcal{M}_1^3(\Delta_y)$, Hermite bicubic) polynomials for concentration. As shown in Figs. 2 and 3, high concentrations of the displacing solvent "finger" toward the production well in a typical displacement, creating a solution after breakthrough that is difficult to approximate with piecewise polynomials on a coarse grid. Along a boundary of the quarter five-spot meeting the production well, the solution is qualitatively as pictured in Fig. 5. Along the diagonal from injector to producer, however, the concentration is close to 1. Allowing themselves on the order of 100 nodes for a quarter five-spot (viewing it as a small part of a larger simulation with dozens of wells), Douglas and Dupont observed a cusping behavior of the C^0 quadratic solution along the boundary for a 5 × 5 grid (Fig. 6). In contrast, the C^1 cubic solution was smeared out (Fig. 7). Grid-orientation sensitivity of oil recovery was noted in both cases; the parallel grid produced more with cubics and the diagonal grid produced more with quadratics. It appeared that a level of smoothness in some sense between C^0 and C^1 would yield a better approximation scheme.

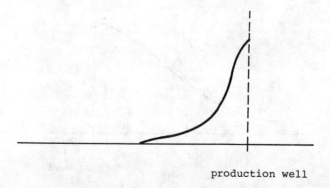

FIG. 5. *Qualitative concentration profile along edge of diagonal grid for five-spot.*

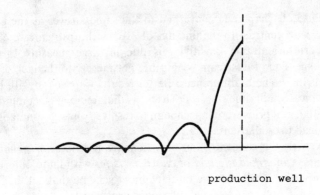

Fig. 6. *Cusping of C^0 quadratic approximation.*

In the calculus of variations, one speaks of adding (positive) penalty terms to a functional in order to bias the minimizing function in some fashion. For the self-adjoint problem (5.1)–(5.2), an equivalent variational form exists as noted in §5, and penalty terms could be added to bias the approximate solution from a Ritz method. For the convection-diffusion equation (6.1) through (6.3), the analogy is no longer perfect, but the same type of process can be carried out. The idea is to measure failure to be C^1 by the size of jumps in normal derivatives across mesh boundaries, and to control these jumps by adding penalty terms containing them to the standard method (6.16).

Let E_0 and E_∂ denote, respectively, the sets of interior and boundary edges in the finite element mesh. Fix a unit normal vector ν on each edge (for a rectangular mesh, use $\hat{e}_x$ or $\hat{e}_y$), taking the outward normal on boundary edges. If $e \in E_0$ and $x \in e$, define the jump of a function f at x across e as

$$[f](x) = \lim_{\varepsilon \downarrow 0} [f(x - \varepsilon\nu(x)) - f(x + \varepsilon\nu(x))].$$

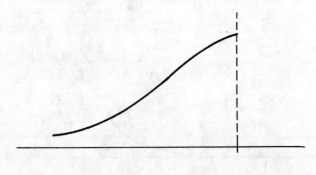

Fig. 7. *Smearing of C^1 cubic approximation.*

For $x \in e \in E_\partial$ let $[f](x) = f(x)$. Now take a nonnegative penalty function $\sigma_1(x, t)$ (usually constant on edges), and add to the left-hand side of (6.16) the term

$$(6.18) \qquad J_1(C^{n+1}, v) = \sum_{e \in E_0} |e| \int_e \sigma_1 \left[\frac{\partial C^{n+1}}{\partial \nu}\right] \left[\frac{\partial v}{\partial \nu}\right] dx.$$

If (6.1) through (6.3) is regarded as a periodic problem with the no-flow boundary condition representing reflection, as is the case for the five-spot miscible displacement problem, then the boundary edges may also be thought of as interior edges with jumps in $\partial C^{n+1}/\partial \nu$ multiplied by 2. In this case, add also

$$(6.19) \qquad J_\partial(C^{n+1}, v) = \sum_{e \in E_\partial} 2|e| \int_e \sigma_1 \left[\frac{\partial C^{n+1}}{\partial \nu}\right] \left[\frac{\partial v}{\partial \nu}\right] dx$$

to (6.16). The edge length $|e|$ appears in (6.18) and (6.19) to scale the resulting matrix terms to the size of those in (6.16); $|e|$ may be regarded as part of σ_1.

Note that $J_1(v, v) \geq 0$ and $J_\partial(v, v) \geq 0$, so the extra positive-semidefinite term does not worsen the behavior of the linear system arising from (6.16). Note also that, if c is smooth (in $L^\infty(J; H^2(\Omega))$, say), then the normal-derivative jumps of c are zero, so that $J_1(c(t), v) = J_\partial(c(t), v) = 0$ for any $v \in \mathcal{M}$. Thus the penalty should not alter the fact that C will converge to c as the mesh and time steps are refined; it will change the nature of C for particular meshes. Douglas and Dupont [23] proved that, if $\mathcal{M} \cap C^1(\Omega)$ is capable of good approximation (e.g., for $\mathcal{M} = \mathcal{M}_0^2(\Delta_x) \times \mathcal{M}_0^2(\Delta_y)$, $\mathcal{M} \cap C^1(\Omega)$ consists of quadratic splines, which can still approximate to order h^3 in $L^2(\Omega)$ and h^2 in $H^1(\Omega)$), then an error estimate like (6.17) holds with K independent of σ_1. Further, the theory demonstrates a bound independent of σ_1 for $J_1(C^{n+1}, C^{n+1})$ and $J_\partial(C^{n+1}, C^{n+1})$, so that as σ_1 is increased on an edge, the numerical normal-derivative jumps on that edge must decrease. By manipulating σ_1, one can fix the smoothness of C^{n+1} anywhere between C^0 ($\sigma_1 = 0$) and C^1 ($\sigma_1 = +\infty$). Numerical computations by Douglas and Dupont with moderate values of σ_1 yielded solutions between the cusped and smeared ones, as desired. Additionally, for reasons that are not intuitively clear, grid-orientation sensitivity was practically eliminated. The penalties were added to the original code by R. E. Bank.

One possible interpretation relates to dispersion. Young [116] theorized that he did not observe grid orientation, even with five-point differences, because unlike previous investigators he used a physical dispersion model. In their original grid-oriented computations without penalties, Douglas and Dupont used a physical *diffusion* model ($d_m > 0$, $d_l = d_t = 0$ in (3.8)). In this model the coefficient does not grow with velocity, so that the Peclet number approaches infinity near wells. Interior penalties introduce mesh-dependent dispersion around a front by forcing the numerical solution toward continuous differentiability when C^1 polynomials may be unable to match the sharpness of the front. Perhaps the penalties added enough dispersion near the production well to make

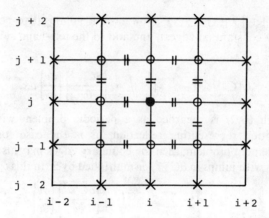

Fig. 8. *Coupling for continuous bilinear interior penalty procedure.*

up for the lack of physical dispersion and the coarseness of the grid. Later computations have shown that the grid used by Douglas and Dupont was too coarse to give good approximation without numerical dispersion.

One must pay a price in computational work for using interior penalties. For simplicity, consider piecewise-bilinear basis functions. Without penalties, (6.16) couples the basis function v_{ij} (1 at (x_i, y_j), 0 at other nodes), represented by the solid dot in Fig. 8, to itself and its eight neighbors represented by circles. Now note that v_{ij} has nonzero normal-derivative jump on all of the edges marked by double lines. Similar considerations for other nodes show that J_1 couples v_{ij} to all basis functions at nodes marked by X's. Effectively, we now have a 21-point scheme instead of a 9-point one. This has serious implications for the amount of space needed to store the resulting system of linear equations in a computer and for the amount of processor time to generate the system and solve it, especially if the system is large and an iterative linear-solution technique is being used. In three dimensions (with penalties on jumps across faces), the coupling goes from 27 nodes to 81.

In addition to starting from C^0 spaces and penalizing normal-derivative jumps to approach C^1, one can consider starting from C^{-1} (discontinuous) spaces and penalizing function jumps to approach C^0. This requires care in reaching an analogue of (6.16), because the discontinuities force us to integrate by parts one element at a time and keep track of the resulting edge integrals. Let T index the elements, with $(f, g)_T = \int_T fg \, dx$. If the average of f at $x \in e$ across edge $e \in E_0$ is

$$\{f\}(x) = \tfrac{1}{2} \lim_{\varepsilon \downarrow 0} [f(x - \varepsilon \nu(x)) + f(x + \varepsilon \nu(x))],$$

then $[fg] = [f]\{g\} + \{f\}[g]$, and (assuming D and u are continuous)

$$-\sum_T (\nabla \cdot (D\nabla C^{n+1} - uC^{n+1}), v)_T$$

$$= \sum_T (\nabla \cdot (D\nabla C^{n+1} - uC^{n+1}), \nabla v)_T$$

(6.20)
$$-\sum_{e \in E_0} \int_e D\left(\left[\frac{\partial C^{n+1}}{\partial v}\right]\{v\} + \left\{\frac{\partial C^{n+1}}{\partial v}\right\}[v]\right) dx$$

$$+ \sum_{e \in E_0} \int_e u \cdot v([C^{n+1}]\{v\} + \{C^{n+1}\}[v]) \, dx.$$

Noting that $[c^{n+1}] = [\partial c^{n+1}/\partial v] = 0$ for the true solution c, we can make the edge terms symmetric by replacing $[\partial C^{n+1}/\partial v]\{v\}$ with $[C^{n+1}]\{\partial v/\partial v\}$, without affecting convergence of C to c. We can define σ_0 and J_0 analogously to σ_1 and J_1, obtaining as our procedure

$$\left(\phi \frac{C^{n+1} - C^n}{\Delta t}, v\right) + \sum_T (D\nabla C^{n+1} - uC^{n+1}, \nabla v)_T$$

(6.21)
$$-\sum_{e \in E_0} \int_e D([C^{n+1}]\left\{\frac{\partial v}{\partial v}\right\} + \left\{\frac{\partial C^{n+1}}{\partial v}\right\}[v]) \, dx$$

$$+ \sum_{e \in E_0} \int_e u \cdot v([C^{n+1}]\{v\} + \{C^{n+1}\}[v]) \, dx$$

$$+ \sum_{e \in E_0} |e|^{-1} \int_e \sigma_0 [C^{n+1}][v] \, dx = (\tilde{C}^{n+1} q^{n+1}, v), \qquad v \in \mathcal{M}.$$

$\mathcal{M}$ could be $\mathcal{M}_{-1}^r(\Delta_x) \otimes \mathcal{M}_{-1}^r(\Delta_y)$ for any $r \geq 1$. The scale factor $|e|^{-1}$ appears in J_0 for the same reason as before. No J_∂ term is needed because $[C^{n+1}] = 0$ on $\partial \Omega$ with a reflection boundary condition.

This type of method was analyzed theoretically for elliptic problems by Wheeler [110] and for parabolic problems by Arnold [3]. Its principal advantage is that, because the basis functions on each element are independent of those on other elements, the choice of mesh is very flexible. For example, two elements may intersect in part of an edge rather than a complete edge, as is usually required. This allows local refinement in a tensor-product mesh without having to extend refining lines across the domain. For dynamic refinement in traveling-front problems, this could be useful. Also, the accumulation matrix in (6.21) is block diagonal, hence is easily invertible; since it is the only term with coefficient $1/\Delta t$, it may be a good approximation of the whole matrix and may serve as an efficient preconditioner for conjugate-gradient solution of the linear equations. The thrust of the theoretical results was that, even on such an irregular mesh, there is a minimum value σ such that for $\sigma_0 \geq \sigma$ an error estimate like (6.17) holds.

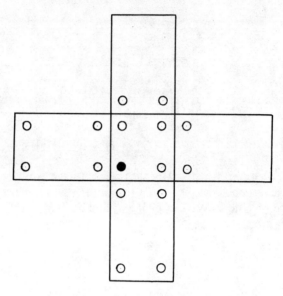

FIG. 9. *Coupling for discontinuous bilinear interior penalty procedure.*

Like its continuous counterpart, this penalty procedure can involve considerable extra computational work. With piecewise-bilinear functions, each element has four basis functions; in the continuous case, only one per node was required. With higher-order polynomials this growth is moderated (e.g., with cubics, from 9 to 16 per node), but in three dimensions it becomes worse. The extra edge integrals demand computer time and programming effort. For bilinears, the function marked by the solid circle in Fig. 9 is coupled to the 16 indicated functions, as opposed to 9 in the continuous case.

By adding J_1 to (6.21), it is possible to combine the two penalty methods. This would allow a discontinuous underlying space, with large σ_0 and moderate σ_1 leading to solutions like those obtained by Douglas and Dupont.

We next describe the modified method of characteristics, which is applied to the nondivergence form (3.14b) of the concentration equation. As in (6.1), assume that u is known. The idea of the method is to time-step along the characteristics of the hyperbolic problem obtained by ignoring diffusion and dispersion in (3.14b). This treats the convection part of the problem, which is the difficult part numerically; then finite differences or finite elements are used for the diffusion-dispersion part, for which they are well-suited.

The time-stepping direction is given by a unit vector $\tau(x)$ defined by

$$(6.22) \qquad \sqrt{\phi(x)^2 + |u(x)|^2}\, \frac{\partial}{\partial \tau} = \phi \frac{\partial}{\partial t} + u_x \frac{\partial}{\partial x} + u_y \frac{\partial}{\partial y}.$$

Using (6.22), we can rewrite (3.14b) in the form

(6.23) $$\sqrt{\phi^2 + |u|^2} \frac{\partial c}{\partial \tau(x)} - \nabla \cdot (D\nabla c) = (\tilde{c} - c)q.$$

Then (6.23), which has the form of a diffusion-dispersion equation, is approximated by backward-differenced finite differences or finite elements.

The crucial aspect of the method, both theoretically and computationally, is the approximation of the "time" derivative $\partial c/\partial \tau$. Many methods based on characteristics fix a point $x \in \Omega$ at time level n and ask where it will go at time level $n + 1$; these *moving point* or *front tracking* methods must then solve at time level $n + 1$ on a mesh of irregular or unpredictable character. In two dimensions this is difficult, and in three it is still much more difficult. Our method takes the opposite view, fixing a point at level $n + 1$ and asking where it came from at level n. Thus the solution mesh at level $n + 1$ is controlled by the method, not the flow; it can be fixed for all time or adjusted to changing flow patterns in an uncomplicated way. Two- and three-dimensional flows present no difficulty in this framework. Our numerical work so far has used fixed mesh.

Accordingly, consider $(x, t^{n+1}) \in \Omega \times J$. If the velocity operating on the fluid particle between t^n and t^{n+1} can be well approximated by $u(x, t^{n+1})$, let $u^* = u(x, t^{n+1})$ and define

(6.24) $$\bar{x} = x - \frac{u^*}{\phi(x)} \Delta t.$$

This approximates the characteristic through (x, t^{n+1}) by its tangent at (x, t^{n+1}). For certain applications, with rapidly varying velocity, u^* will have to be chosen more carefully. For a function f, set $\bar{f}(x) = f(\bar{x})$. Then form the backward-difference approximation

$$\frac{\partial c^{n+1}}{\partial \tau}(x) \approx \frac{c^{n+1}(x) - c^n(\bar{x})}{\Delta t \sqrt{1 + \left|\frac{u^*}{\phi}\right|^2}},$$

which is equivalent to

(6.25) $$\sqrt{\phi^2 + |u^*|^2} \frac{\partial c^{n+1}}{\partial \tau}(x) \approx \phi \frac{c^{n+1}(x) - \bar{c}^n(x)}{\Delta t}.$$

Thus, a finite difference approximation of (6.23) can take the form

(6.26) $$\phi \frac{C_{ij}^{n+1} - \bar{C}_{ij}^n}{\Delta t} - \{\text{difference formula for } \nabla \cdot (D\nabla c)_{ij}^{n+1}\} = (\tilde{C} - C)_{ij}^{n+1} q_{ij}^{n+1},$$

where the diffusive-dispersive difference formula is necessarily a nine-point scheme if D is a full-tensor physical dispersion model. Analogously, a finite element scheme is

(6.27) $$\left(\phi \frac{C^{n+1} - \bar{C}^n}{\Delta t}, v\right) + (D\nabla C^{n+1}, \nabla v) = ((\tilde{C} - C)^{n+1} q^{n+1}, v), \quad v \in \mathcal{M}.$$

Note that in either method, the convection term moves to the right-hand side in $\overline{C}^n$, so that the system of linear equations is symmetric and positive-definite. This makes the method ideal for iterative linear-solution algorithms in large problems. In (6.26), $\overline{C}^n_{ij} = \overline{C^n((x_i, y_j))}$ does not, in general, lie at a grid point, so some type of interpolation is necessary to evaluate it. The finite element case requires no interpolation since the elements already do it, but the integral $(\phi \overline{C}^n, v)$ must be evaluated by a suitable quadrature rule. The choice of this rule is not obvious, because $\overline{C}^n$ is not the type of function considered by the usual theory; it will have corners at points interior to elements that are mapped back to element edges by approximate characteristics. Because of this, we have used Lobatto rules in our numerical work, desiring information from every element in the characteristic image of a given element. In either method, approximate characteristics may cross $\partial\Omega$; if they do, the no-flow boundary condition can be used as a reflection to continue C.

With the above notation, we can see that this method falls into the class of fractional-step procedures, with the method of characteristics taking C^n to $\overline{C}^n$, then finite differences or finite elements taking $\overline{C}^n$ to C^{n+1}. The characteristic step does not conserve mass, and we shall say more about this in §7. Characteristics are not new to the petroleum literature; Garder, Peaceman, and Pozzi [37] developed a moving-point method for miscible displacement almost 20 years ago, but as we noted above, our procedure is quite different.

For a single equation like (6.1) with smooth data, Douglas and Russell [27] analyzed this procedure theoretically. For finite differences on nonuniform grids with linear interpolation for $\overline{C}^n$, they found the expected error of order $h + \Delta t$ at the grid points. For uniform grids, they were able to replace h by h^2 if a form of quadratic interpolation was used. For the finite element method, they obtained an optimal estimate like (6.17). For both methods, the usual $\|\partial^2 c/\partial t^2\|\Delta t$ temporal error estimate was replaced by $\|\partial^2 c/\partial \tau^2\|\Delta t$, which should be much smaller because the true solution changes much less rapidly along characteristics that follow the flow than it does at a fixed point in space. The Galerkin proof did not need the techniques for treating nonsymmetric terms that inflate the constant in (6.17). Thus, there is reason to expect much larger time steps to be feasible and accurate with this method than with standard schemes. It was noted that the theory extended to nonlinear problems. Ewing and Russell [30] extended the finite element theory of Douglas and Russell [37] to a three-level second-order backward-difference method, obtaining the same optimal spatial error together with the expected $\|\partial^3 c/\partial \tau^3\|(\Delta t)^2$ error in time. Their techniques could be extended further to higher-order backward-difference methods. High-order time stepping methods do not make sense for traveling-front problems in conjunction with standard procedures, because the time derivatives of the solution are large. However, by time stepping with the flow, we can make high-order methods worthwhile. Pironneau [77] analyzed essentially the same single-step finite element scheme for the Navier–Stokes equations; his proof obtained one order of accuracy less in space than the proof in [27], but he allowed the analogue of diffusion and dispersion to go down to zero.

Numerical results for single parabolic equations in one dimension have been reported by Ewing and Russell [30] and Russell [82]. Neuman [61] used the same ideas in numerical computations for contaminant transport. A constant-coefficient convection-diffusion equation was approximated by single-step and multistep methods [30], and the nonlinear Burgers' equation was treated by single-step schemes [82]. In each case an exact solution was available to allow comparison of errors for various methods. If the spatial mesh was too coarse to resolve a front, characteristics did not improve the solution much. Once the grid was fine enough (perhaps three intervals across a front), characteristics enabled much larger time steps to be used with no loss of accuracy. In some cases, time steps could be two orders of magnitude larger than in standard methods. The answers with characteristics were also remarkably free of oscillations and numerical dispersion. The multistep method improved over the single-step scheme when the mesh was very fine, so as to practically eliminate spatial error. With Burgers' equation, no accuracy was lost if the nonlinearity was lagged along characteristics, avoiding a system of nonlinear equations at each time step; this gives hope that the method may be able to handle more complicated nonlinearities in the same way. The straight characteristics of these example equations allowed the method to conserve mass; we shall say more about its nonconservativeness for miscible displacement in §7.

The cell-balance finite element procedure was introduced by Potempa [78]. It is a hybrid of ideas from block-centered finite differences and piecewise-bilinear finite elements. In §5, we saw that block-centered differences follow from conservation of mass on each grid block in a domain. Now consider the variational form (6.15) of the concentration equation, without Green's theorem:

$$(6.28) \quad \left(\phi \frac{\partial c}{\partial t}, v\right) - (\nabla \cdot (D\nabla c - uc), v) = (\tilde{c}q, v), \quad t \in J, \quad v \in H^1(\Omega).$$

Actually, (6.28) holds for $v \in L^2(\Omega)$, and if we choose v to be the characteristic function χ_{ij} of grid block $[x_{i-1}, x_i] \times [y_{j-1}, y_j] = R_{ij}$, (6.28) becomes precisely the statement of mass conservation for R_{ij}. In the finite element method, we do not use χ_{ij} as test function, but rather (for example) a continuous piecewise bilinear function v_{ij} that is 1 at (x_i, y_j) and zero at all other nodes. Since v_{ij} is supported on $\Omega_{ij} = R_{ij} \cup R_{i,j+1} \cup R_{i+1,j} \cup R_{i+1,j+1}$, we can regard (6.28) with $v = v_{ij}$ as a weighted mass balance on Ω_{ij}. Potempa considers the region in $\mathbb{R}^3$ under the graph of v_{ij} and views (6.28) with $v = v_{ij}$ as a mass balance on that region. We shall think instead of a mass balance on $\Omega_{ij} \subset \mathbb{R}^2$ with measure $v_{ij}(x, y) dx\, dy$. Potempa's method mimics block-centered differences, but uses the regions Ω_{ij} with weighted measures rather than the standard grid blocks; it is written as a finite difference scheme.

To mimic block-centered differences, we must define the mass in a region Ω_{ij}

and the rate of flow (using a Darcy velocity) from one weighted region to another. For incompressible miscible displacement, using volume instead of mass, the volume of the solvent component in Ω_{ij} is

$$\text{(6.29)} \qquad \int_{y_{j-1}}^{y_{j+1}} \int_{x_{i-1}}^{x_{i+1}} \phi c v_{ij} \, dx \, dy.$$

For the inter-region flow let X_i and Y_j be the piecewise-linear functions such that $v_{ij}(x, y) = X_i(x) Y_j(y)$. Consider an area element $dx\, dy$ in $R_{i+1,j+1}$, and analyze flow in the x-direction from Ω_{ij} to $\Omega_{i+1,j}$ and $\Omega_{i+1,j+1}$ (no x-direction flow from Ω_{ij} to $\Omega_{i,j+1}$) under velocity u_x. The interstitial velocity is u_x/ϕ, and in time $dx/(u_x/\phi)$ this area element will lose its mass to a like element to its right. This mass (volume) of *all* components is $\phi\, dx\, dy$, and its weight in Ω_{ij} will change by $\partial v_{ij}/\partial x \, dx$. The proportion Y_j of this flow will go into $\Omega_{i+1,j}$, and the remaining proportion $1 - Y_j = Y_{j+1}$ will go into $\Omega_{i+1,j+1}$. The flow rate from Ω_{ij} to $\Omega_{i+1,j}$ due to the element $dx\, dy$ is

$$\frac{(\phi\, dx\, dy)\,((\partial v_{ij}/\partial x)\, dx)\, Y_j}{dx/(u_x/\phi)} = u_x \frac{\partial v_{ij}}{\partial x} Y_j \, dx \, dy.$$

Thus, the total flow rate is

$$\text{(6.30)} \qquad \Gamma_{ij}^{i+1,j} = \iint_{R_{i+1,j} \cup R_{i+1,j+1}} u_x \frac{\partial v_{ij}}{\partial x} Y_j \, dx \, dy.$$

Similarly,

$$\text{(6.31)} \qquad \Gamma_{ij}^{i+1,j+1} = \iint_{R_{i+1,j+1}} \left(u_x \frac{\partial v_{ij}}{\partial x} Y_{j+1} + u_y \frac{\partial v_{ij}}{\partial y} X_{i+1} \right) dx\, dy.$$

Rates of flow to the other six neighbors of (x_i, y_j) are obtained by symmetry from (6.30) or (6.31). Set Γ_{ij}^{ij} to zero. These quantities, multiplied by a time step Δt, are analogous to V_R and V_L of the derivation of block-centered differences in §5. Potempa then sets up an upstream-weighted finite difference scheme along the lines of (6.4), of the form (mass balance for solvent component only)

$$\text{(6.32)} \qquad \left(\iint \phi v_{ij}\, dx\, dy \right) \frac{C_{i,j}^{n+1} - C_{i,j}^n}{\Delta t}$$

$$= - \sum_{k=-1}^{+1} \sum_{l=-1}^{+1} \Gamma_{i,j}^{i+k,j+l} C_{i+k/2, j+l/2}^{n+1} + \iint v_{ij} \tilde{C}^{n+1} q^{n+1} \, dx\, dy,$$

where

$$\text{(6.33)} \qquad C_{i+k/2, j+l/2} = \begin{cases} C_{i,j} & \text{if } \Gamma_{i,j}^{i+k,j+l} > 0, \\ C_{i+k, j+l} & \text{if } \Gamma_{i,j}^{i+k,j+l} < 0. \end{cases}$$

Because the accumulation term is lumped, this scheme can be made explicit also.

From the way the method was derived, it is clear that it conserves mass. Additionally, Bell, Shubin, and Wheeler [5] have shown that the implicit version satisfies a maximum principle. Thus, it will have no problem with overshoot, and likely will not experience any nonphysical oscillations. In fact, if u_x and u_y are constant the procedure reduces to the nine-point scheme of Yanosik and McCracken [114] described earlier. The reduced grid-orientation sensitivity and considerable numerical dispersion of that procedure are shared by Potempa's method; the leading part of its numerical dispersion term is (6.13).

Potempa's philosophy in developing this procedure was to reduce grid orientation within a framework easily implemented in existing petroleum industry simulators. The method does this, because it differs from that of Yanosik and McCracken only in its use of velocity integrals, rather than pressure differences, to determine flow coefficients. These integrals are easily calculated by inexpensive quadrature rules; for example, u can be evaluated at the midpoint of a cell and the rest of the integral evaluated exactly. It is interesting that Potempa's flow concept extends to nonrectangular meshes, triangles, for example. This has enabled computations to be done for the seven-spot miscible displacement shown

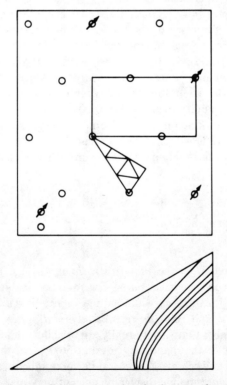

FIG. 10. *Finite element cell-balance procedure for the inverted seven-spot pattern.*

in Fig. 10. For highly nonuniform meshes, results have not been as satisfactory as for uniform grids, probably because of the nature of (6.13) in that case; thus, the method is presently not suitable for accurate simulation of displacements with high Peclet number because of its numerical dispersion (see (6.10)). It should be possible to modify it for better treatment of nonuniform meshes.

The implicit version of the method has been proved to converge for smooth coefficients to the solution of (6.1) by Bell, Shubin, and Wheeler [5], provided that a physical diffusion term is incorporated in (6.32). The error estimate is $\max_n \| c^n - C^n \| = O(h + \Delta t)$, where C^n is bilinearly interpolated to a function. It follows from this work that the method of Yanosik and McCracken, and any other method with a similar type of mesh-dependent diffusion and a maximum principle, is also convergent.

7. Theoretical and numerical results for the miscible displacement system. In this section we shall describe results, both theoretical and computational, for various combinations of the techniques defined in §§5 and 6 for the miscible displacement problem. We begin by specifying test problems that have been the focus of most of the experimental work.

Figure 3 of Chapter I displays a five-spot pattern of injection and production wells. Assume that the reservoir is horizontal, of thickness 1 foot (otherwise, divide the three-dimensional quantities by the thickness to obtain a two-dimensional model), and initially saturated with oil ($c_0(x) \equiv 0$ in (3.13)). Let the distance between an injector and the nearest producers be 1000 $\sqrt{2}$ feet, so that the diagonal grid is a 1000-foot square. Impose a constant flow rate of 200 ft^3/day on each well, a uniform porosity of 0.10 on the reservoir, and a viscosity of 1 centipoise on the oil. Each injector governs a 2000-ft-square five-spot of pore volume $(0.10)(2000)^2(1) = 400{,}000$ ft^3, so that a pore volume is injected in 2000 days. The viscosity of the fluid mixture is assumed to obey the quarter-power law

$$\frac{1}{\mu^{1/4}} = \frac{c_{\text{oil}}}{\mu_{\text{oil}}^{1/4}} + \frac{c_{\text{solvent}}}{\mu_{\text{solvent}}^{1/4}},$$

which can be written in the form

$$\mu(c) = [(1 - c) + M^{1/4}c]^{-4},$$

where $c = c_{\text{solvent}}$ and M is the *mobility ratio* $\mu_{\text{oil}}/\mu_{\text{solvent}}$. The mobility ratio, diffusion-dispersion coefficients d_m, d_l, and d_t, and permeability k will vary from experiment to experiment. If $d_m = 0$, then the longitudinal and transverse Peclet numbers are $1000/d_l$ and $1000/d_t$, respectively. If $d_l = d_t = 0$, there is a single isotropic Peclet number that varies areally, approaching infinity at wells, zero at the nonwell corners of the diagonal grid, and $56/d_m\phi$ at the center. We shall follow Young [116] and refer to the central Peclet number in such cases.

For $M > 1$, the more mobile displacing solvent may make multiple unstable channels into the oil, in the phenomenon of viscous fingering. The greater the value of M, the more unstable the displacement front. Even if the front is stable, greater values of M cause the solvent to preferentially move toward the production well, reducing oil recovery; see, for example, [28]. As noted in §2, dispersion and heterogeneity are also controlling factors here. It is fairly clear heuristically, and presumably could be made rigorous, that the mathematical solution of a homogeneous problem should not have multiple fingers; the point on the front directly between the injector and the producer is closest to the producer, so should move the fastest, and other points should move more slowly as one traces away from the fastest point. Nonzero dispersion should make the problem well-posed in the sense of continuous dependence on data, for arbitrarily high mobility ratio. Thus, we take the view that any multiple fingers observed in simulation of homogeneous problems are due to numerical instabilities and not to the modeling of physics. We shall be seeking single, stable fingers in homogeneous problems and multiple fingers in heterogeneous problems with adverse mobility ratio; those fingers *will* be due to modeling of physics.

In addition to viscous fingering, our numerical studies will be concerned with grid-orientation sensitivity, numerical dispersion, overshoot and undershoot (or lack thereof as dictated by maximum principles), and material-balance errors (or conservation of mass). Traditionally, grid orientation has been measured by differences in recovery curves (e.g., Fig. 5 of Chapter I) for diagonal and parallel grids. We feel that this is not adequate, because heterogeneities may make the location of concentration level curves in the reservoir (e.g., Fig. 4 of Chapter I) just as important as outlet concentration at a production well. Hence, we present areal concentration maps along with recovery curves. Numerical dispersion is evaluated rather subjectively by comparing different methods for the same problem and by varying grid spacing and time-step size. Overshoot (undershoot) is the greatest amount by which concentration exceeds 1 (is less than 0) at some suitably chosen time. Schemes based on the divergence form (3.10b) of the concentration equation should conserve mass; taking $v \equiv 1$ in (6.16) yields

$$(7.1) \qquad \int_\Omega \phi C^{n+1} dx = \int_\Omega \phi C^n dx + \Delta t \int_\Omega \tilde{C}^{n+1} q^{n+1} dx.$$

For the nondivergence form (3.14b), one obtains

$$(7.2) \qquad \int_\Omega \phi C^{n+1} dx = \int_\Omega \phi C^n dx + \Delta t \int_\Omega [(\tilde{C} - C)^{n+1} q^{n+1} - u \cdot \nabla C^{n+1}] dx.$$

If $\nabla \cdot u = q$ and $u \cdot v = 0$, as is the case if u is obtained from a mixed method, then

$$(7.3) \qquad \begin{aligned} -\int_\Omega u \cdot \nabla C^{n+1} dx &= \int_\Omega [-\nabla \cdot (uC^{n+1}) + C^{n+1}(\nabla \cdot u)] \, dx \\ &= -\int_{\partial\Omega} C^{n+1}(u \cdot v) ds + \int_\Omega C^{n+1} q^{n+1} \, dx \\ &= \int_\Omega C^{n+1} q^{n+1} \, dx, \end{aligned}$$

and (7.2) reduces to (7.1). If u is obtained otherwise, then the nondivergence form does not conserve mass, and material-balance errors will be used as a criterion of acceptability of numerical solutions. The modified method of characteristics defined in (6.27) is nonconservative even if u comes from a mixed method, because the convection term in (7.2) is handled by approximations to characteristics instead of a variational integral.

Overshoot and undershoot have appeared in the early stages of displacements, when the front is nearly radial around the injection well (assuming uniform, isotropic permeability) and very steep (see Fig. 1). Methods without excessive numerical dispersion have difficulty representing such fronts in reasonable cartesian grids. If M is large, these numerical errors may lead to nonphysical viscous fingers. To alleviate these problems, many of our experiments have assumed radial flow in the early stages, solving a one-dimensional equation cheaply until the front is far enough away from the injector to be representable in a two-dimensional mesh. We derive here the equation that has been used. With incompressible radial flow, the Darcy velocity is $\tilde{u}(r) = q/2\pi r$, so that no pressure equation is needed. There is no transverse dispersion, so we have $D = \phi d_m + d_l \tilde{u} = \phi d_m + d_l q / 2\pi r$. Consider an annulus A around the injection well of inner radius r_0 and outer radius r_1. An equation balancing accumulation, convection, diffusion, and dispersion of solvent on A (no sources) is

$$\text{(7.4)} \quad \frac{\partial}{\partial t}\left(\int_A \phi c \, dx\right) = qc(r_0) - qc(r_1) - 2\pi r_0 D \frac{\partial c}{\partial r}(r_0) + 2\pi r_1 D \frac{\partial c}{\partial r}(r_1)$$

$$= \int_{\partial A} \left(D \frac{\partial c}{\partial r} - \tilde{u} c\right) \cdot \nu \, ds.$$

Using the polar-coordinate representation $\nabla \cdot f = (1/r)(\partial/\partial r)(rf(r))$ of the divergence operator, we can use the divergence theorem in (7.4), rewrite in polar coordinates, and drop integrations to obtain

$$\text{(7.5)} \quad 2\pi r \phi \frac{\partial c}{\partial t} - \frac{\partial}{\partial r}\left(2\pi r D \frac{\partial c}{\partial r} - qc\right) = 0, \quad r > r_{\text{well}}, \quad t > 0.$$

The boundary condition at the injection well is an influx of $\tilde{c}q$, which is represented by

$$\text{(7.6)} \quad 2\pi r_{\text{well}} D \frac{\partial c}{\partial r}(r_{\text{well}}) - qc(r_{\text{well}}) = -\tilde{c}q,$$

and we can require $c = 0$ at some distance R far from the well. The initial condition is complete saturation with oil. We can handle the point-source case by letting $r_{\text{well}} \to 0$, which yields

$$\text{(7.7a)} \quad 2\pi r \phi \frac{\partial c}{\partial t} - \frac{\partial}{\partial r}\left([2\pi r \phi d_m + d_l q] \frac{\partial c}{\partial r} - qc\right) = 0, \quad r > 0, \quad t > 0,$$

(7.7b) $\quad c - d_l \dfrac{\partial c}{\partial r} = \tilde{c}, \quad r = 0, \quad t > 0,$

(7.7c) $\quad c = 0, \quad r = R, \quad t > 0,$

(7.7d) $\quad c = 0, \quad r > 0, \quad t = 0.$

It is a simple matter to solve (7.7) by finite elements, using the Neumann condition (7.7b) when integrating by parts. Note that, because of dispersion at the well, the flowing concentration $\tilde{c}$ and the resident concentration c may differ; this can happen at production wells also, but our experiments have neglected it so far.

The theoretical results for finite elements will refer to spaces of piecewise polynomials of degree r and s for the concentration and pressure approximations, respectively. These will be defined on meshes of maximum diameter h_c and h_p. The proofs of the theorems in the cited references simplify to a single parameter h; it is routine to specialize the arguments to obtain estimates like $O(h_c^{r+1} + h_p^{s+1})$ from $O(h^{r+1} + h^{s+1})$, and we make no further note of this here. As noted at the end of §4, it may be worthwhile to use different time steps $\Delta t_p > \Delta t_c$; this will be mentioned where appropriate. All of the theorems, unless noted otherwise, assume that the source and sink term q is smoothly distributed rather than a combination of Dirac measures. This lays a firm foundation for the methods and should demonstrate, as indicated by the numerical results of [34], the expected rates of convergence away from wells. Around wells, these rates will not be observed in practice.

The remainder of this section is organized by concentration method. Within each concentration method, we consider various pressure methods and the theoretical and numerical results associated with each combination. The concentration methods, with pressure methods in parentheses, are five-point finite differences (five-point finite differences), the finite element cell-balance scheme of Potempa (standard Galerkin), a standard Galerkin method (standard Galerkin, mixed), continuous interior penalties (standard Galerkin), discontinuous interior penalties (standard Galerkin, mixed), and the modified method of characteristics with finite elements (standard Galerkin, mixed) and nine-point finite differences (five-point finite differences, mixed). The combined time stepping is in the manner of (4.6) unless otherwise indicated.

The five-point difference scheme, with upstream weighting in the concentration equation, is a combination of (5.3) (block-centered grid) and (6.4). This would be the sequential method used in the petroleum industry and is included 1ere for completeness. We know of no theoretical convergence results for the coupled system; indeed, there is evidence that, at least at practical levels of discretization, it fails to converge to the same answer for different grid orientations [16]. For constant permeability, $M = 100$, and $d_m = d_l = d_t = 0$ (numerical dispersion only), Fig. 11 displays diagonal and parallel concentration maps and

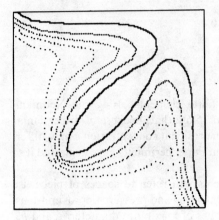
Concentration at .5 pore volume, diagonal grid, 20 × 20 mesh.

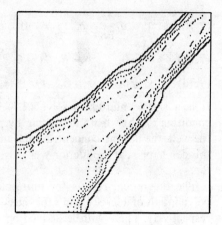
Concentration at .5 pore volume, parallel grid, 28 × 28 mesh.

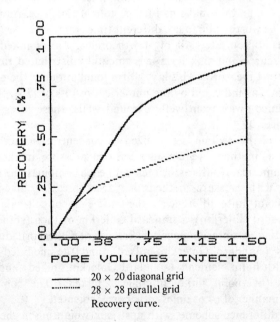
—— 20 × 20 diagonal grid
······ 28 × 28 parallel grid
Recovery curve.

FIG. 11. *Five-point upstream-weighting finite difference procedure for the five-spot pattern with* $M = 100$.

recovery curves. The fronts are considerably smeared, and the sensitivity to grid orientation is clear. The method does conserve mass, and because of the upstream weighting it should be easy to show that it satisfies a maximum principle, so that overshoot and undershoot do not occur.

Analogous results are shown in Fig. 12 for the finite element cell-balance

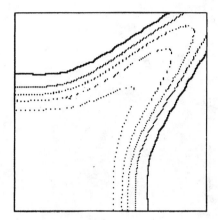
Concentration at .5 pore volume, diagonal grid, 20 × 20 mesh.

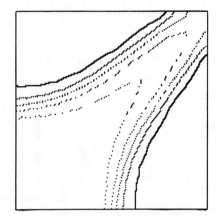
Concentration at .5 pore volume, parallel grid, 28 × 28 mesh.

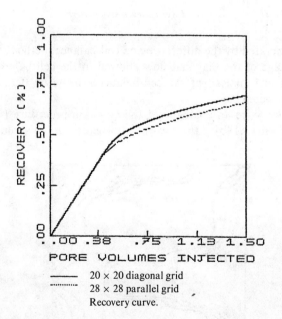
——— 20 × 20 diagonal grid
·········· 28 × 28 parallel grid
Recovery curve.

FIG. 12. *Finite element cell-balance (Potempa) for the five-spot pattern with $M = 100$.*

procedure (6.32) with Galerkin pressure (5.4). This shares the virtues of the difference method and, as expected from §6, reduces grid orientation because its nine-point dispersion from upstream weighting is nearly rotationally invariant (see (6.14)). Its Peclet number, like that of the difference method, is estimated by (6.10). Fig. 13 shows an inverted seven-spot pattern with no-flow boundaries, a problem that permits one to study grid orientation in a single plot. Figures 14

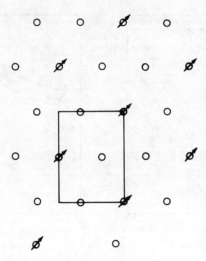

FIG. 13. *Inverted seven-spot pattern.*

and 15 were generated by the difference and cell-balance methods, respectively, and the advantage of the diagonal flow allowed in the cell-balance scheme is evident. Theoretical analysis of the cell-balance scheme by Bell, Shubin, and Wheeler [5] was noted in §6.

One can easily write an analogue of the Galerkin procedure (6.16) for the nondivergence form (3.14b) of the concentration equation. This, coupled to (5.4),

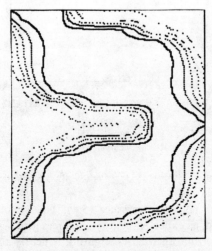

FIG. 14. *Concentration at .5 pore volume (five-point upstream-weighting finite difference procedure for the inverted seven-spot pattern, 19×22 grid, $M = 100$).*

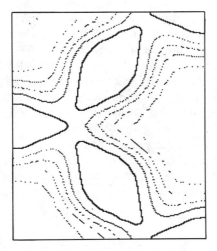

FIG. 15. *Concentration at .5 pore volume (finite element cell-balance procedure for the inverted seven-spot pattern,* 19×22 *grid,* $M = 100$).

was analyzed by Ewing and Wheeler [32]. For a continuous-time version, in which the time derivative in (6.15) is not replaced by a difference quotient, they proved that

$$\|C - c\|_{L^\infty(J;L^2(\Omega))} + h\|\nabla(P - p)\|_{L^\infty(J;L^2(\Omega))} + h\|\nabla(C - c)\|_{L^2(J;L^2(\Omega))}$$

(7.8)
$$= \begin{cases} O(h^{r+1} + h^{s+1} + h^{r+s-1}) & \text{if } d_l = d_t = 0, \quad r + s \geq 3, \\ O(h^{r+1} + h^s) & \text{if } d_l \neq 0 \text{ or } d_t \neq 0, \quad s \geq 2. \end{cases}$$

Convergence for piecewise-linear polynomials for both pressure and concentration was not demonstrated. They also analyzed discrete-time schemes, obtaining estimates like (7.8) with additional error terms of Δt for backward differencing and $(\Delta t)^2$ for Crank–Nicolson.

Analyses of efficient time-stepping algorithms for the standard Galerkin approach can be found in Ewing and Russell [31]. In particular, they investigated the use of different time steps for pressure and concentration, with Δt_p an integer multiple of Δt_c and velocity extrapolated as in (4.6). For backward differencing, they found time-truncation errors of $O(\Delta t_c + (\Delta t_p)^2)$ in addition to the spatial errors of (7.8), so that one should choose $\Delta t_p = O((\Delta t_c)^{1/2})$. The use of preconditioned conjugate gradient iteration for the systems of linear equations, iterating only long enough to stabilize the time-stepping procedure, was considered in this paper; the idea had been discussed for a single nonlinear parabolic equation in [24]. In certain cases, Ewing and Russell showed that a fixed number of iterations per time step, independent of h and Δt, was sufficient for stability and optimal convergence. Their arguments also lead to an improved estimate in

(7.8); the term h^{r+s-1} can be replaced by $h^2|\log h|$, if $d_l = d_t = 0$ and $r = s = 1$. This convergence of linear polynomials for both unknowns has not been proved for the more physical case of tensor dispersion.

Numerical results with standard Galerkin methods were cited in §6 as motivation for the interior-penalty procedures. Cusping (a form of overshoot and undershoot) and grid orientation made the results unsatisfactory. These results used Dirac measures as source terms at wells and subtracted out pressure singularities as in (5.12). Ewing and Wheeler [33] were able to analyze this theoretically for the restricted case of $M = 1$ (unit mobility ratio), in which viscosity is independent of concentration, pressure does not change with time, and (3.14b) becomes a linear parabolic equation with singular source. Their results were, for continuous time and any $\varepsilon > 0$,

$$\|\tilde{u} - U\| = O(h^{1-\varepsilon}),$$

(7.9)
$$\|c - C\|_{L^\infty(J;L^2(\Omega))} = \begin{cases} O(h^{1-\varepsilon}) & \text{if } d_l = d_t = 0, \\ O(h^{1/2-\varepsilon}) & \text{if } d_l \neq 0 \text{ or } d_t \neq 0. \end{cases}$$

These theorems depend on regularity results of Sammon [85], who proved that for any $\varepsilon > 0$ and $r < \infty$,

(7.10)
$$\|c\|_{L^2(J;H^{2-\varepsilon}(\Omega))} + \|c\|_{L^2(J;W^1_r(\Omega))} + \|c\|_{L^\infty(J;H^{1-\varepsilon}(\Omega))} + \|p\|_{L^\infty(J;H^{1-\varepsilon}(\Omega))}$$
$$+ \|\tilde{u}\|_{L^\infty(J;L^{2-\varepsilon}(\Omega))} + \left\|\frac{\partial c}{\partial t}\right\|_{L^2(J;L^{2-\varepsilon}(\Omega))} \leq K.$$

The nondivergence analogue of (6.16) coupled to the mixed method (5.18)–(5.19) was analyzed by Douglas, Ewing, and Wheeler [25]. With smooth source and continuous time, they derived optimal rates of convergence for concentration, pressure, and velocity:

(7.11)
$$\|c - C\|_{L^\infty(J;L^2(\Omega))} = O(h^{r+1} + h^{s+1}),$$
$$\|\tilde{u} - U\|_{L^\infty(J;V)} + \|p - P\|_{L^\infty(J;W)} = O(h^{r+1} + h^{s+1}),$$

where V and W are as in §5 and the spaces W^s_h and V^s_h of (5.20) are used. For singular source and $M = 1$, they obtained $\|\tilde{u} - U\| = O(h|\log h|)$, together with concentration estimates as in (7.9). In another paper [26], they considered an efficient time-stepping procedure for this combination of methods. This procedure takes Δt_p to be an integer multiple of Δt_c, but instead of extrapolating velocity as in (4.6) it extrapolates to the middle of the current pressure time step. Compensating terms involving the difference between the two extrapolations are added to the right-hand side, with concentration replaced by an extrapolation. The result is that over a pressure time step, the concentration matrix remains constant; it can be factored once by a direct method and used several times. This efficient direct method is an alternative to the iterative methods analyzed by

Ewing and Russell [31]. Time-truncation errors of $O(\Delta t_c + (\Delta t_p)^2)$ were obtained in addition to the spatial errors of (7.11). No numerical experiments have been conducted with this particular combination.

The continuous interior-penalty Galerkin procedure (nondivergence analogue of (6.16), with (6.18) and (6.19) added) for concentration was analyzed by Wheeler and Darlow [111] in combination with a standard Galerkin method (5.4) for pressure. Continuous-time and discrete-time cases were considered, in particular, the use of different time steps for pressure and concentration. The analysis, which combines ideas of Douglas and Dupont [23] and of Ewing and Wheeler [32], involves a restriction on the time rate of change of σ_1 in (6.18) and (6.19); it suffices to assume that there exist constants $\sigma^* > 0$ and K such that

$$(7.12) \quad h\left[\sup\left\{\left|\frac{\partial \sigma_1}{\partial t}(x,t)\right| : 0 \leq \sigma_1(x,t) \leq \sigma^*\right\} + \sup\left\{\sigma_1(x,t)^{-1}\left|\frac{\partial \sigma_1}{\partial t}(x,t)\right| : \sigma_1(x,t) \geq \sigma^*\right\}\right] \leq K.$$

An estimate of the form of (7.8) for $d_l = d_t = 0$ was demonstrated, with appropriate time-truncation errors for the discrete-time schemes. Numerical results with subtraction of singularities were cited in §6; the behavior of standard Galerkin methods was improved upon significantly.

The discontinuous interior-penalty method (6.21), analyzed for a single parabolic equation by Arnold [3] as noted in §6, has not been analyzed as part of a coupled system. Numerical results with the standard Galerkin method (5.4) for pressure (with singularities removed) have been reported by Darlow [17] and by Douglas, Wheeler, Darlow, and Kendall [28]. In the latter paper, mobility ratios of 1, 10, and 41 were used, with $\phi d_m = 1$ ft^2/day, $d_l = d_t = 0$ (Peclet number 56), and various permeability patterns. In accordance with theory, it was found that σ_0 (actually $\sigma_0 d_m$) had to exceed a minimum value to stabilize the scheme; values of the order of 100 ft^2/day were found to make the approximate solution nearly continuous. If σ_1 was also made large, the piecewise-bilinear solution became nearly C^1; thus, any edge with large σ_1 was effectively removed from the mesh, and self-adaptivity concepts could be tested in a comparatively simple program with a fixed mesh. On edges to be kept in the grid, best results were obtained with $\sigma_1 = 0.5$ ft^2/day. As one would expect, solutions were insensitive to removal of edges in parts of the grid away from the front. The study used the nondivergence form of the concentration equation, concentration time steps of 8 days (0.004 pore volumes injected) with (effectively) an underlying uniform 20×20 grid, and radial initialization (7.7) to 0.1 pore volumes injected. The efficient time-stepping scheme analyzed in Douglas, Ewing, and Wheeler [26] was also used. For uniform permeability, material-balance errors were of the order of 1%, an acceptable level, and grid orientation (as measured by recovery curves) was very slight. The penalty σ_1 introduces some numerical dispersion, not clearly understood. Overshoot and undershoot were significant (approximately 9%), even for this rather modest Peclet number. Runs were made with nonuniform permeability, and satisfactory results were obtained with variations up to 10 to 1 (1000 millidarcies in half of the domain, 100 in the other half). However, the method did not perform well for variations of 100 to 1, and convergence problems were encountered with a mobility ratio of 100 [17].

Some of these difficulties were alleviated by using the mixed method (5.18)–(5.19), with velocity singularities removed as in (5.26) through (5.28), in place of the standard Galerkin method for pressure. These calculations by Darlow, Ewing, and Wheeler [18] used the conservative divergence form and bilinear polynomials in (6.21), with $r = 1$ in (5.20). Figure 16 presents diagonal and parallel concentration maps and recovery curves for $M = 100$, $\phi d_m = 1$ ft^2/day, $d_l = d_t = 0$ (Peclet number 56), and uniform permeability of 100 md. Radial initialization was used to 0.1 pore volumes injected. Material-balance errors of about 0.5% arose from inexact solution of the linear equations by preconditioned conjugate gradient iteration. Grid orientation is clearly absent, but numerical dispersion and overshoot/undershoot are as serious as for the Galerkin pressure. The gains from the mixed method were the ability to handle higher mobility ratio, as shown here, and greater variations in permeability; the latter is to be expected, since one no longer has to multiply by the rough permeability coefficient to obtain velocity.

The overshoots in these methods are not instabilities, because they do not grow with time. They are reflections of the inability of a particular grid to resolve a moving front without additional numerical dispersion. The modified method of characteristics (6.27) reduces these difficulties by, in effect, stopping the front; the question of overshoot comes down to how well the front can be projected into a piecewise-polynomial space. Coarse grids will still cause overshoot, but it will not be as serious as in the standard Galerkin and interior-penalty schemes.

The combination of the finite element modified method of characteristics with a standard Galerkin procedure for pressure was analyzed by Russell [80]. Results of the form of (7.8) were derived, with time-truncation error of $O(\Delta t)$ depending on the characteristic derivative $\partial^2 c/\partial \tau^2$ as in §6. Numerical results were reported by Russell [81], using biquadratic polynomials with singularities removed for pressure and bilinears for concentration. The code was actually based on discontinuous bilinears with penalties, but the penalties were not helpful and were suppressed (i.e., σ_0 was made large and σ_1 was set to zero); effectively, it was an inefficient continuous bilinear code. Radial initialization was used to 0.01 pore volumes injected, with sufficiently fine grid around injection wells to handle the front thereafter. Because of the smaller time-truncation error, large time steps of 40 to 80 days (0.02 to 0.04 pore volumes) could be taken without loss of accuracy, as was demonstrated by varying h and Δt in different runs.

The accumulation term $(\phi \bar{C}^n, v)$ in (6.27) is evaluated by numerical integration; as noted in §6, Lobatto rules (usually Simpson's rule) were used. An

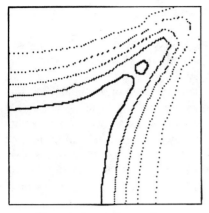

Concentration at .5 pore volume,
diagonal grid, 10 × 10 mesh.

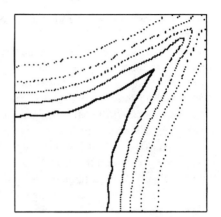

Concentration at .5 pore volume,
parallel grid, 14 × 14 mesh.

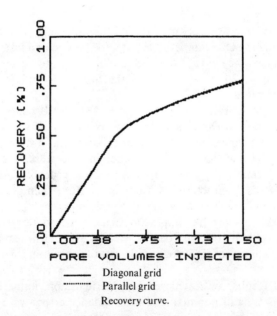

——— Diagonal grid
·········· Parallel grid
Recovery curve.

FIG. 16. *Interior penalty Galerkin procedure for the five-spot pattern with $M = 100$.*

exception was made in corner cells containing wells, because it is easier if no quadrature point falls at a well. For points near wells, u^* in (6.24) must be computed carefully because the velocity field (extrapolated from two previous pressure time steps) changes rapidly in space. In [81], an attempt was made to do this semi-analytically by using the knowledge of the singular behavior of the velocity; this proved to be unsatisfactory and was replaced in the later numerical work of Ewing, Russell, and Wheeler [29].

The results, all with uniform permeability, showed very little grid orientation, numerical dispersion (as seen by varying h and Δt), or overshoot (0.5% for $M = 100$, $\phi d_m = 1$, $d_l = d_t = 0$ on a 20×20 diagonal grid). A problem with $M = 41$ and longitudinal and transverse Peclet numbers of 100 and 1000 was handled successfully. Material-balance errors were another matter. While generally around 3%, they ranged as high as 10% in certain cases. Most of these errors occurred after solvent breakthrough at the production wells, indicating that the semi-analytic method for tracing characteristics from quadrature points near production wells was inadequate. Indeed, take $v \equiv 1$ in (6.27) and assume that $\tilde{C} = C$ at wells (this is true at production wells, and also at injectors after a short time). The result is

$$(7.13) \qquad \int_\Omega \phi C^{n+1}(x)\, dx = \int_\Omega \phi C^n(\bar{x})\, dx,$$

which shows that mass-conservation error is determined entirely by how accurately one follows the characteristics of the velocity field and integrates $\phi \bar{C}^n$, and by the accuracy of the velocity field itself. This motivated a better scheme for computing u^* in (6.24), and suggested the mixed method (5.18)–(5.19) for pressure and velocity.

It should be straightforward to analyze (5.18)–(5.19) coupled to (6.27) by combining the results of Douglas, Ewing, and Wheeler [25] with those of Russell [80]; this has not been carried out. Numerical computations have been reported by Ewing, Russell, and Wheeler [29], combining the pressure code of Darlow, Ewing, and Wheeler [18] and the concentration code of Russell [81] with penalties removed. Characteristics from quadrature points near wells were traced in segments corresponding to micro-timesteps, each of these involving a trapezoidal predictor-corrector approximation of a chord along the characteristic. The size of each micro-timestep was selected by an error estimator using the inverse-radius behavior of the velocity field; details appear in Ewing, Russell, and Wheeler [29].

The desired results were achieved—grid orientation, numerical dispersion, and overshoot were still minimal, and mass-balance errors were reduced to the range of 0.5% in most runs. Figure 17 shows diagonal and parallel concentration maps and recovery curves for uniform permeability, $M = 100$, $\phi d_m = 1$ ft^2/day, $d_l = 0$ foot, and $d_t = 0$ foot. With the mixed method, it was possible to model random permeability variations of six orders of magnitude, as is shown by the physical fingers in Fig. 18 for the same mobility ratio and dispersion coefficients. The same problem was simulated by the cell-balance scheme (6.32) with physical dispersion replaced by numerical dispersion, with results pictured in Fig. 19. It is clear that the cell-balance method is much more dispersive for this example.

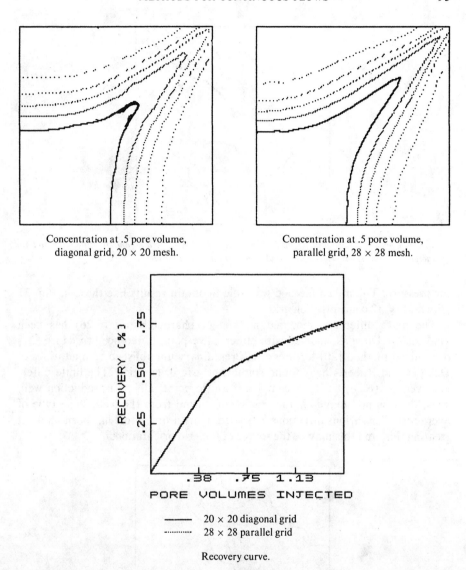

FIG. 17. *Modified method of characteristics for the five-spot pattern with $M = 100$.*

The mixed method also improved the stability of the procedure, helping to suppress nonphysical fingers. Figures 20 and 21 show concentration maps for $M = 41$, uniform permeability, $\phi d_m = 0$, $d_l = 10$ feet, and $d_t = 1$ foot. These were computed by the finite element modified method of characteristics for concentration on 32 × 32 and 64 × 64 grids, respectively, and a standard Galerkin method

FIG. 18. *Concentration solution at .3 pore volume, mixed method and modified method of characteristics—random permeability and M = 100 (five-spot pattern).*

for pressure. The mixed method was able to obtain results like those in Fig. 21 with a 32 × 32 concentration grid.

The finite difference modified method of characteristics (6.26) has been analyzed by Douglas, coupled with either a five-point difference procedure [21] or a mixed method [20] for pressure. Preliminary unpublished computations of Douglas and Roberts have found significant grid orientation. The finite difference version (6.26) becomes complex if a grid point falls at a production well, since there is no unique characteristic emanating from the well; some type of averaging over various directions is needed. Only a first effort has been made at treating this, and this may be the source of the grid orientation.

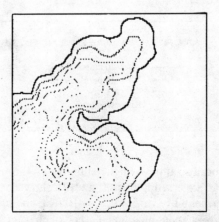

FIG. 19. *Concentration solution at .3 pore volume, finite element cell-balance procedure—random permeability and M = 100 (five-spot pattern).*

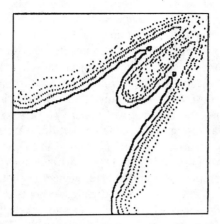

FIG. 20. *Concentration solution at .5 pore volume, M = 41, 32 × 32 concentration grid.*

8. Future directions of research. The research reported to date has answered many questions, but it poses many others. We suggest possible lines of future investigation here. We consider ideas applicable to the various methods introduced in §§5 and 6, and then discuss more general ideas.

In the pressure methods, we need better treatment of near-well singularities (near-singularities in the true physical problems), both theoretically and computationally. All of our high-order convergence theorems assume that the source and sink terms of (3.10) or (3.14) are bounded; for singular terms, the only proofs to date are for the unphysical assumption of concentration-independent viscosity. This theory needs to be extended to the physical case. Computationally, removal

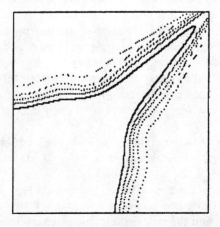

FIG. 21. *Concentration solution at .5 pore volume, M = 41, 64 × 64 concentration grid.*

of singularities works well if the coefficient $a = k/\mu$ is constant, in which case the singularity can be removed from pressure, or if flow is radial, allowing removal from the velocity in the mixed method. A sensible procedure that has not been tried is to remove from velocity around an injection well, where flow is radial if permeability is constant and isotropic, and from pressure around a production well, where flow is not radial after solvent breakthrough. In cases of high mobility ratio, the viscosity contrast will make k/μ vary by direction from the producer, and singularity removal becomes more difficult; perhaps a directional dependence should be incorporated into the theory. The methods should be easily extendable to anisotropic permeability (k_x, k_y distinct in (3.2) of Chapter I) by replacing circles by ellipses. For highly heterogeneous problems, the subtracted singularities may have to be cut off in small neighborhoods of the wells.

The mixed method demands better techniques for solving its system of linear equations. So far we have used a conjugate gradient algorithm preconditioned by the diagonal of the matrix A of (5.32). This has been quite slow and better preconditioners are needed. The analogy between block-centered differences and mixed methods has suggested the use of the block-centered difference matrix as a preconditioner, and early results show satisfactory convergence in approximately ten iterations, independent of mesh size and heterogeneity. Another possibility is to use an alternating-direction iteration as a preconditioner either for the mixed method itself [11] or for the block-centered preconditioner.

The modified method of characteristics for concentration needs accurate velocities in order to maintain small mass-conservation error, motivating the use of mixed methods for pressure. It would be of interest to carry out experiments on how accurate these velocities need to be and what order mixed method is most efficient for obtaining them. We would like to see performance comparisons between the lowest-order method ($r = 0$) on a fine grid and the next ($r = 1$) on a coarser grid. The question of conservation for the modified method of characteristics is an overriding one, and we should consider ways of post-processing the procedure to make it conservative. The conservation error in a typical time step is of the order of 10^{-4}, so it should be possible via some projection or extrapolation of the numerical solution to compensate for this.

The mixed method allows easy treatment of random permeability variations down to grid level. Since the interaction of heterogeneity, dispersion, mobility ratio, and instability (viscous fingering) is of such interest as noted in §2, we could fix a permeability distribution on a moderately coarse grid and then observe the behavior of the solution as the grid is refined. This refinement might reach levels requiring considerable expense, but it could shed a great deal of light on the fundamentals of the problem, perhaps showing the way for subsequent theoretical work. As an example, consider the plots in Figs. 20 and 21, generated for mobility ratio 41 and homogeneous permeability with a standard Galerkin method for pressure and the modified method of characteristics for the concentration. The multiple-fingered case used a 32 × 32 mesh for concentration, while the stable case used 64 × 64. Otherwise the two runs were the same. Because the modified method of characteristics avoids numerical dispersion, oscillation, and grid orientation on reasonable grids, it should be possible to model the physics with a fair degree of accuracy.

Another convergence study is suggested by some fragmentary numerical results with mixed methods for a single elliptic equation. These have pointed to possible superconvergence (convergence at a higher rate than the possible global order of approximation) of velocities at Gauss nodal points. This type of convergence is well known for Galerkin procedures at element nodes, and theoretical and numerical investigations for mixed methods would be worthwhile.

Many workers in this area have thought that it would be interesting to somehow combine finite elements near wells with finite differences elsewhere in the reservoir, because finite differences are faster but finite elements are better equipped to handle complicated near-well flow. Potempa's finite difference method based on finite element cells might provide ideas in this direction. There is no reason why that procedure demands upstream weighting; its numerical dispersion could be curtailed by use of midpoint weighting combined with a physical dispersion model discretized in the usual ways. The idea of flow between finite element cells would allow one to set up a global finite difference scheme, with standard block-centered flow coefficients away from wells and finite-element-cell flows near wells.

A somewhat neglected area in our numerical work is that of time-step selection. We have found that the modified method of characteristics permits much larger time steps than other procedures, but this has not been systematically quantified. It would be useful to develop automatic time-step selectors that would push each method to its limits and/or balance spatial and temporal errors.

Analogously, in the space variables adaptive grid refinement or generation should be considered. The advantages of the modified method of characteristics do not appear unless the grid is fine enough to resolve fronts, and in large-scale problems this may require adaptive mesh.

Finally, the procedures will have to be extended to more complex problems than two-component single-phase displacement. The mixed method will be more complicated for multiphase flow because each phase has its own pressure and Darcy velocity; ideas like those of Chavent et al. [12] are one way of handling this. It should be possible to extend the modified method of characteristics to compressible, multiphase, multicomponent flows by noting that a fluid particle in a phase moves according to the interstitial velocity of that phase, which is its Darcy velocity (including the relative permeability as a multiplier) divided by the product of its saturation and the rock porosity. This allows transport of component compositions by the method of characteristics, after which phase behavior (mass transfer) could be updated and dispersion incorporated via finite differences or finite elements.

9. Conclusions. This research was motivated by the significant difficulties encountered by standard finite difference and finite element methods in simulating enhanced recovery processes of petroleum engineering. These processes lead to convection-dominated parabolic partial differential equations strongly coupled to elliptic or parabolic equations, as exemplified by the two-component model of miscible displacement. Finite difference procedures used in the petroleum industry smear traveling concentration fronts far beyond the level of physical dispersion, and replacement of upstream weighting by midpoint weighting would cause unacceptable nonphysical oscillations. Answers can be severely dependent on the orientation of the computational grid. Standard Galerkin methods can exhibit the same grid-orientation sensitivity for high-mobility-ratio displacements on coarse grids, yield nonphysical cusp-like solutions on coarse grids, and oscillate seriously even on rather fine grids. Dominant convection forces all of these methods to use small time steps.

The interior-penalty Galerkin procedures developed to attack grid orientation and cusping and to provide flexibility in grid refinement appear now to be too cumbersome and costly for the benefits they provide. The discontinuous version increases the order of the system of linear equations needed to attain a given accuracy of approximation, and the continuous one increases the bandwidth and density of nonzero entries in this system. Time steps must still be small, penalties on normal-derivative jumps add numerical dispersion, and even on rather fine grids solutions show overshoot and oscillations. Grid orientation can be alleviated by virtually any finite element method or non-upstream-weighted finite difference method on meshes adequate to resolve traveling fronts, and similarly cusping appears to have been an indication that the mesh was too coarse to handle the problem in question without numerical dispersion. If numerical dispersion is needed, it can be added more cheaply and without grid orientation by increasing the physical dispersion coefficient. The flexibility in grid refinement offered by the discontinuous penalty procedure may be advantageous in certain cases, but the authors have found the method cumbersome to program and expensive to run, and it would seem that there must be easier ways to achieve the same gains.

The mixed method for pressure and velocity looks very promising if better ways of solving the associated system of linear equations can be found. The scheme treats highly heterogeneous problems, with permeabilities varying by six orders of magnitude, as easily as homogeneous problems. It has been shown theoretically and numerically to calculate velocities more accurately than the usual procedures that compute pressure and then difference or differentiate.

The modified method of characteristics for concentration also looks very promising, provided that mass conservation can be kept under control and that the scheme can be generalized to more complex problems. Mass conservation is dependent on accurate velocities, making the coupling of this procedure with mixed methods particularly attractive. Theoretically and numerically, the method has been shown to lose no accuracy with time steps much larger than those of other methods, perhaps one to two orders of magnitude larger, provided that the spatial grid is fine enough to resolve fronts. Answers show essentially no grid orientation, numerical dispersion, or overshoot. For problems with very high Peclet number, say in the thousands, adaptive grid refinement will be needed on the current generation of computers.

Methods with mesh-dependent numerical dispersion, such as the upstream finite difference method based on finite element cell balances, are too dispersive for accurate simulation of problems with Peclet number in the hundreds unless adaptive refinement is used. They can provide qualitative results in an efficient and uncomplicated manner. The concept of finite element cell balances extends to meshes of arbitrary geometry.

Overall, we feel that significant progress has been made in attacking the numerical difficulties typical of realistic petroleum reservoir simulation. The authors and others will continue these investigations and attempt to bring the ideas of this paper closer to practical application.

Acknowledgments. A great many of the ideas in this paper are due, in one form or another, to Jim Douglas, Jr. His influence on both of the authors' views of reservoir simulation problems has been considerable. The assistance of T. C. Potempa in writing and running a block-centered finite difference code, and in producing plots with his finite element cell-balance procedure, was very valuable. We are indebted to Amoco Production Co., Arco Oil and Gas Co., Exxon Production Research Co., Getty Oil Co., IBM Corp., Marathon Oil Co., and Mobil Research and Development Corp., which supported much of this research. The first author thanks R. L. Christiansen and H. Kazemi for helpful discussions and the management of Marathon Oil Company for permission to publish this paper. The second author was supported in part by the National Science Foundation under grant MCS 80-03025.

REFERENCES

[1] R. A. ADAMS, *Sobolev Spaces*, Academic Press, New York, 1975.
[2] J. O. AMAEFULE AND L. L. HANDY, *The effect of interfacial tensions on relative oil/water permeabilities of consolidated porous media*, Soc. Pet. Eng. J., 22 (1982), pp. 371–381.
[3] D. N. ARNOLD, *An interior penalty finite element method with discontinuous elements*, Ph.D. thesis, Univ. Chicago, 1979; SIAM J. Numer. Anal., 19 (1982), pp. 742–760.
[4] K. AZIZ AND A. SETTARI, *Petroleum Reservoir Simulation*, Applied Science Publishers, London, 1979.
[5] J. BELL, G. SHUBIN, AND M. F. WHEELER, *Analysis of a new method for computing the flow of miscible fluids in a porous medium*, to appear.
[6] R. J. BLACKWELL, J. R. RAYNE, AND W. M. TERRY, *Factors influencing the efficiency of miscible displacement*, Trans. AIME, 216 (1959), pp. 1–8.

[7] J. H. BRAMBLE, *On the convergence of difference approximations for second order uniformly elliptic operators,* in Numerical Solution of Field Problems in Continuum Physics, Vol. II, American Mathematical Society, Providence, RI, 1970, pp. 201–209.

[8] J. H. BRAMBLE, B. E. HUBBARD, AND V. THOMÉE, *Convergence estimates for essentially positive type discrete Dirichlet problems,* Math. Comp., 23 (1969), pp. 695–710.

[9] F. BREZZI, *On the existence, uniqueness and approximation of saddle-point problems arising from Lagrangian multipliers,* RAIRO Anal. Numer., 2 (1974), pp. 129–151.

[10] W. E. BRIGHAM, P. W. REED, AND J. N. DEW, *Experiments on mixing during miscible displacement in porous media,* Soc. Pet. Eng. J., 1 (1961), pp. 1–8.

[11] D. C. BROWN, *Alternating-direction iterative schemes for mixed finite element methods for second order elliptic problems,* Ph.D. thesis, Univ. Chicago, 1982.

[12] G. CHAVENT, J. JAFFRE, G. COHEN, M. DUPUY, AND I. DIESTE, *Simulation of two-dimensional waterflooding using mixed finite element methods,* SPE 10502, 6th SPE Symposium on Reservoir Simulation, New Orleans, 1982, pp. 147–158.

[13] R. L. CHUOKE, P. VAN MEURS, AND C. VAN DER POEL, *The instability of slow, immiscible, viscous liquid-liquid displacements in permeable media,* Trans. AIME, 216 (1959), pp. 188–194.

[14] P. G. CIARLET, *The Finite Element Method for Elliptic Problems,* North-Holland, Amsterdam, 1978.

[15] E. L. CLARIDGE, *Discussion of the use of capillary tube networks in reservoir performance studies,* Soc. Pet. Eng. J., 12 (1972), pp. 352–361.

[16] K. H. COATS, W. D. GEORGE, AND B. E. MARCUM, *Three-dimensional simulation of steamflooding,* Soc. Pet. Eng. J., 14 (1974), pp. 573–592.

[17] B. L. DARLOW, *A penalty-Galerkin method for solving the miscible displacement problem,* Ph.D. thesis, Rice Univ., Houston, 1980.

[18] B. L. DARLOW, R. E. EWING, AND M. F. WHEELER, *Mixed finite element methods for miscible displacement problems in porous media,* SPE 10501, 6th SPE Symposium on Reservoir Simulation, New Orleans, 1982, pp. 137–145; Soc. Pet. Eng. J., to appear.

[19] J. DOUGLAS JR., *The numerical solution of miscible displacement in porous media,* in Computational Methods in Nonlinear Mechanics, J. T. Oden, ed., North-Holland, Amsterdam, 1980, pp. 225–238.

[20] ———, *Simulation of miscible displacement in porous media by a modified method of characteristic procedure,* Numerical Analysis, Dundee 1981, Lecture Notes in Mathematics 912, Springer-Verlag, Berlin, 1982.

[21] ———, *Finite difference methods for two-phase incompressible flow in porous media,* SIAM J. Numer. Anal., 20 (1983), pp. 681–696.

[22] J. DOUGLAS JR. AND T. DUPONT, *Galerkin methods for parabolic equations,* SIAM J. Numer. Anal., 7 (1970), pp. 575–626.

[23] ———, *Interior penalty procedures for elliptic and parabolic Galerkin methods,* Computing Methods in Applied Science, Lecture Notes in Physics 58, Springer-Verlag, Berlin, 1976.

[24] J. DOUGLAS JR., T. DUPONT, AND R. EWING, *Incomplete iteration for time-stepping a Galerkin method for a quasilinear parabolic problem,* SIAM J. Numer. Anal., 16 (1979), pp. 503–522.

[25] J. DOUGLAS JR., R. E. EWING, AND M. F. WHEELER, *Approximation of the pressure by a mixed method in the simulation of miscible displacement,* RAIRO Anal. Numer., 17 (1983), pp. 17–33.

[26] ———, *A time-discretization procedure for a mixed finite element approximation of miscible displacement in porous media,* RAIRO Anal. Numer., to appear.

[27] J. DOUGLAS JR. AND T. F. RUSSELL, *Numerical methods for convection-dominated diffusion problems based on combining the method of characteristics with finite element or finite difference procedures,* SIAM J. Numer. Anal., 19 (1982), pp. 871–885.

[28] J. DOUGLAS JR., M. F. WHEELER, B. L. DARLOW, AND R. P. KENDALL, *Self-adaptive finite element simulation of miscible displacement in porous media*, SIAM J. Sci. Statist. Comput. to appear.

[29] R. E. EWING, T. F. RUSSELL, AND M. F. WHEELER, *Simulation of miscible displacement using mixed methods and a modified method of characteristics*, SPE 12241, 7th SPE Symposium on Reservoir Simulation, San Francisco, Nov. 1983.

[30] R. E. EWING AND T. F. RUSSELL, *Multistep Galerkin methods along characteristics for convection-diffusion problems*, in Advances in Computer Methods for Partial Differential Equations IV, R. Vichnevetsky and R. S. Stepleman, eds., IMACS, Rutgers Univ., New Brunswick, NJ, 1981, pp. 28–36.

[31] ———, *Efficient time-stepping procedures for miscible displacement problems in porous media*, SIAM J. Numer. Anal., 19 (1982), pp. 1–67.

[32] R. E. EWING AND M. F. WHEELER, *Galerkin methods for miscible displacement problems in porous media*, SIAM J. Numer. Anal., 17 (1980), pp. 351–365.

[33] ———, *Galerkin methods for miscible displacement problems with point sources and sinks—unit mobility ratio case*, Proc. Special Year in Numerical Analysis, Lecture Notes #20, Univ. Maryland, College Park, 1981, pp. 151–174.

[34] ———, *Computational aspects of mixed finite element methods*, in IMACS Transactions on Scientific Computation, Vol. 1, R. S. Stepleman, ed., North-Holland, Amsterdam, 1983.

[35] R. S. FALK AND J. E. OSBORN, *Error estimates for mixed methods*, RAIRO Anal. Numer., 14 (1980), pp. 249–277.

[36] T. C. FRICK, ed., *Petroleum Production Handbook*, McGraw-Hill, New York, 1962, pp. 23–36.

[37] A. O. GARDER, D. W. PEACEMAN, AND A. L. POZZI, *Numerical calculation of multidimensional miscible displacement by the method of characteristics*, Soc. Pet. Eng. J., 4 (1964), pp. 26–36.

[38] S. GERSCHGORIN, *Fehlerabschätzung für das Differenzenverfahren zur Lösung partieller Differentialgleichungen*, Z. Angew. Math. Mech., 10 (1930), pp. 373–382.

[39] W. B. GOGARTY, *Status of surfactant or micellar methods*, J. Pet. Tech., 28 (1976), pp. 93–102.

[40] S. B. GORELL AND G. M. HOMSY, *A theory of the optimal policy of oil recovery by secondary displacement processes*, SIAM J. Appl. Math., 43 (1983), pp. 79–88.

[41] W. G. GRAY, *A derivation of the equations for multiphase transport*, Chem. Eng. Sci., 30 (1975), pp. 229–233.

[42] W. G. GRAY AND K. O'NEILL, *On the general equations for flow in porous media and their reduction to Darcy's law*, Water Resources Res., 12 (1976), pp. 148–154.

[43] S. P. GUPTA, *Dispersive mixing effects on the Sloss field micellar system*, Soc. Pet. Eng. J., 22 (1982), pp. 481–492.

[44] B. HABERMANN, *The efficiency of miscible displacement as a function of mobility ratio*, Trans. AIME, 219 (1960), pp. 264–272.

[45] L. L. HANDY, *An evaluation of diffusion effects in miscible displacement*, Trans. AIME, 216 (1959), pp. 382–384.

[46] L. J. HAYES, R. P. KENDALL, AND M. F. WHEELER, *The treatment of sources and sinks in steady-state reservoir engineering simulations*, in Advances in Computer Methods for Partial Differential Equations II, R. Vichnevetsky, ed. IMACS, Rutgers Univ., New Brunswick, NJ, 1977, pp. 301–306.

[47] J. P. HELLER, *The interpretation of model experiments for the displacement of fluids through porous media*, AIChE J., 9 (1963), pp. 452–459.

[48] ———, *Onset of instability patterns between miscible fluids in porous media*, J. Appl. Phys., 37 (1966), pp. 1566–1579.

[49] T. J. R. HUGHES, ed., *Finite Element Methods for Convection Dominated Flows*, ASME, New York, 1979.

[50] H. KAZEMI, *Low-permeability gas sands,* J. Pet. Tech., 34 (1982), pp. 2229–2232.
[51] R. B. KELLOGG, *An error estimate for elliptic difference equations on a convex polygonal,* SIAM J. Numer. Anal., 3 (1966), pp. 79–90.
[52] H. A. KOCH JR. AND R. L. SLOBOD, *Miscible slug process,* Trans. AIME, 210 (1957), pp. 40–47.
[53] H.-O. KREISS, T. A. MANTEUFFEL, B. K. SWARTZ, B. WENDROFF, AND A. B. WHITE, *More aspects of supratruncation error convergence on nonuniform meshes,* presented at SIAM National Meeting, Denver, June 1983.
[54] C. R. KYLE AND R. L. PERRINE, *Experimental studies of miscible displacement instability,* Soc. Pet. Eng. J., 5 (1965), pp. 189–195.
[55] J. W. LACEY, J. E. FARIS, AND F. H. BRINKMAN, *Effect of bank size on oil recovery in the high-pressure gas-driven LPG-bank process,* J. Pet. Tech., 13 (1961), pp. 806–816.
[56] L. W. LAKE AND F. HELFFERICH, *Cation exchange in chemical flooding: Part 2—The effect of dispersion, cation exchange, and polymer/surfactant adsorption on chemical flood environment,* Soc. Pet. Eng. J., 18 (1978), pp. 435–444.
[57] L. W. LAKE AND G. J. HIRASAKI, *Taylor's dispersion in stratified porous media,* Soc. Pet. Eng. J., 21 (1981), pp. 459–468.
[58] A. LALLEMAND-BARRES AND P. PEAUDECERF, *Investigation of the relationship between the value of the macroscopic dispersiveness of an aquifer medium, its other characteristics and the measurement conditions—bibliographic study,* Bull. BRGM, Ser. 2, 3, #4 (1978), pp. 277–284.
[59] M. C. LEVERETT, *Flow of oil-water mixtures through unconsolidated sands,* Trans. AIME, 132 (1939), pp. 149–171.
[60] T. A. MANTEUFFEL AND A. B. WHITE, presentation, SIAM National Meeting, Denver, June 1983.
[61] S. P. NEUMAN, *An Eulerian-Lagrangian numerical scheme for the dispersion-convection equation using conjugate space-time grids,* J. Comp. Phys., 41 (1981), pp. 270–294.
[62] J. A. NITSCHE AND J. C. C. NITSCHE, *Error estimates for the numerical solution equations,* Arch. Rational Mech. Anal., 5 (1960), pp. 293–306.
[63] J. T. ODEN AND J. N. REDDY, *An Introduction to the Mathematical Theory of Finite Elements,* John Wiley, New York, 1976.
[64] H. D. OUTMANS, *Nonlinear theory for frontal stability and viscous fingering in porous media,* Soc. Pet. Eng. J., 2 (1962), pp. 165–176.
[65] J. G. PATEL, M. G. HEGDE, AND J. C. SLATTERY, *Further discussion of two-phase flow in porous media,* AIChE J., 18 (1972), pp. 1062–1063.
[66] D. W. PEACEMAN, *Improved treatment of dispersion in numerical calculation of multidimensional miscible displacements,* Soc. Pet. Eng. J., 6 (1966), pp. 213–216.
[67] ———, *Fundamentals of Numerical Reservoir Simulation,* Elsevier, Amsterdam, 1977.
[68] ———, *Interpretation of well-block pressures in numerical reservoir simulation,* SPE 6893, 52nd Annual Fall Technical Conference and Exhibition, Denver, 1977; Soc. Pet. Eng. J., 18 (1978), pp. 183–194.
[69] ———, *Interpretation of well-block pressures in numerical reservoir simulation with nonsquare grid blocks and anisotropic permeability,* Soc. Pet. Eng. J., 23 (1983), pp. 531–543.
[70] J. PEAZ, P. REED, AND J. C. CALHOUN, *Relationship between oil recovery, interfacial tension, pressure gradient in water-wet porous media,* Prod. Mon., 19 (1955), No. 7, pp. 34–38.
[71] T. K. PERKINS AND O. C. JOHNSTON, *A review of diffusion and dispersion in porous media,* Soc. Pet. Eng. J., 3 (1963), pp. 70–84.
[72] ———, *A study of immiscible fingering in linear models,* Soc. Pet. Eng. J., 9 (1969), pp. 39–46.
[73] T. K. PERKINS, O. C. JOHNSTON, AND R. N. HOFFMAN, *Mechanics of viscous fingering in miscible systems,* Soc. Pet. Eng. J., 5 (1965), pp. 301–317.

[74] R. L. PERRINE, *The development of stability theory for miscible liquid-liquid displacement*, Soc. Pet. Eng. J., 1 (1961), pp. 17–25.
[75] ——, *A unified theory for stable and unstable miscible displacement*, Soc. Pet. Eng. J., 3 (1963), pp. 205–213.
[76] E. J. PETERS AND D. L. FLOCK, *The onset of instability during two-phase immiscible displacement in porous media*, Soc. Pet. Eng. J., 21 (1981), pp. 249–258.
[77] O. PIRONNEAU, *On the transport-diffusion algorithm and its application to the Navier-Stokes equations*, Numer. Math., 38 (1982), pp. 309–332.
[78] T. C. POTEMPA, *Finite element methods for convection dominated transport problems*, Ph.D. thesis, Rice Univ., Houston, 1982.
[79] P. A. RAVIART AND J. M. THOMAS, *A mixed finite element method for 2nd order elliptic problems*, in Mathematical Aspects of the Finite Element Method, Rome 1975, Lecture Notes in Mathematics, Springer-Verlag, Berlin, 1977.
[80] T. F. RUSSELL, *An incompletely iterated characteristic finite element method for a miscible displacement problem*, Ph.D. thesis, Univ. Chicago, Chicago, 1980.
[81] ——, *Finite elements with characteristics for two-component incompressible miscible displacement*, SPE 10500, 6th SPE Symposium on Reservoir Simulation, New Orleans, 1982, pp. 123–135.
[82] ——, *Galerkin time stepping along characteristics for Burgers' equation*, in IMACS Transactions on Scientific Computation, Vol. 1, R. S. Stepleman, ed., North-Holland, Amsterdam, 1983, pp. 183–192.
[83] P. G. SAFFMAN AND G. I. TAYLOR, *The penetration of a fluid into a porous medium or Hele-Shaw cell containing a more viscous liquid*, Proc. Roy. Soc., A245 (1958), pp. 312–329.
[84] A. A. SAMARSKII, *Local one dimensional schemes on non-uniform nets*, Zh. Vych. Nat., 3, 3 (1963), pp. 431–466. (In Russian.)
[85] P. H. SAMMON, *Numerical approximations for a miscible displacement process in porous media*, to appear.
[86] A. SETTARI AND K. AZIZ, *Use of irregular grid in reservoir simulation*, Soc. Pet. Eng. J., 12 (1972), pp. 103–114.
[87] ——, *Use of irregular grid in cylindrical coordinates*, Soc. Pet. Eng. J., 14 (1974), pp. 396–404; Discussion, pp. 405–408.
[88] A. SETTARI, H. S. PRICE AND T. DUPONT, *Development and application of variational methods for simulation of miscible displacement in porous media*, Soc. Pet. Eng. J., 17 (1977), pp. 228–246.
[89] J. C. SLATTERY, *Flow of viscoelastic fluids through porous media*, AIChE J., 13 (1967), pp. 1066–1071.
[90] ——, *Multiphase viscoelastic flow through porous media*, AIChE J., 14 (1968), pp. 50–56.
[91] ——, *Single-phase flow through porous media*, AIChE J., 15 (1969), pp. 866–872.
[92] ——, *Two-phase flow through porous media*, AIChE J., 16 (1970), pp. 345–352.
[93] ——, *Interfacial effects in the entrapment and displacement of residual oil*, AIChE J., 20 (1974), pp. 1145–1154.
[94] ——, *Interfacial effects in the recovery of residual oil by displacement*, in 1982 report to our industrial consortium, Dept. of Chemical Engineering, Northwestern Univ., Evanston, IL, 1982.
[95] R. L. SLOBOD AND S. J. LESTZ, *Use of a graded viscosity zone to reduce fingering in miscible phase displacements*, Prod. Mon., 24 (1960), No. 10, pp. 12–19.
[96] R. L. SLOBOD AND R. A. THOMAS, *Effect of transverse diffusion on fingering in miscible-phase displacement*, Soc. Pet. Eng. J., 3 (1963), pp. 9–13.
[97] G. STRANG AND G. J. FIX, *An Analysis of the Finite Element Method*, Prentice-Hall, Englewood Cliffs, NJ, 1973.

[98] J. J. TABER, *Dynamic and static forces required to move a discontinuous oil phase from porous media containing both oil and water*, Soc. Pet. Eng. J., 9 (1969), pp. 3–12.
[99] G. I. TAYLOR, *Dispersion of soluble matter in solvent flowing slowly through a tube*, Proc. Roy. Soc., A219 (1953), pp. 186–203.
[100] ———, *Conditions under which dispersion of a solute in a stream of solvent can be used to measure molecular diffusion*, Proc. Roy. Soc., A225 (1954), pp. 473–477.
[101] M. R. TODD, P. M. O'DELL, AND G. J. HIRASAKI, *Methods for increased accuracy in numerical reservoir simulators*, Soc. Pet. Eng. J., 12 (1972), pp. 515–530.
[102] G. W. THOMAS, *Principles of Hydrocarbon Reservoir Simulation*, International Human Resources Development Corp., Boston, 1982.
[103] C. VAN DER POEL, *Effect of lateral diffusivity on miscible displacement in horizontal reservoirs*, Soc. Pet. Eng. J., 2 (1962), pp. 317–326.
[104] O. R. WAGNER AND R. O. LEACH, *Effect of interfacial tension on displacement efficiency*, Soc. Pet. Eng. J., 6 (1966), pp. 335–344.
[105] J. E. WARREN AND J. C. CALHOUN JR., *A study of waterflood efficiency in oil-wet systems*, Trans. AIME, 204 (1955), pp. 22–29.
[106] J. E. WARREN AND F. F. SKIBA, *Macroscopic dispersion*, Soc. Pet. Eng. J., 4 (1964), pp. 215–230.
[107] J. W. WATTS AND W. J. SILLIMAN, *Numerical dispersion and the origins of grid orientation effect—a summary*, Paper 96C, presented at the 73rd Annual Meeting of AIChE, Chicago, 1980.
[108] A. WEISER AND M. F. WHEELER, *Block centered finite difference methods*, to appear.
[109] M. F. WHEELER, *A priori L_2 error estimates for Galerkin approximations to parabolic partial differential equations*, SIAM J. Numer. Anal., 10 (1973), pp. 723–759.
[110] ———, *An elliptic collocation finite element method with interior penalties*, SIAM J. Numer. Anal., 15 (1978), pp. 152–161.
[111] M. F. WHEELER AND B. L. DARLOW, *Interior penalty Galerkin methods for miscible displacement problems in porous media*, in Computational Methods in Nonlinear Mechanics, J. T. Oden, ed., North-Holland, Amsterdam, 1980, pp. 458–506.
[112] S. WHITAKER, *Diffusion and dispersion in porous media*, AIChE J., 13 (1967), pp. 420–427.
[113] ———, *Advances in theory of fluid motion in porous media*, Ind. Eng. Chem., 61 (1969), No. 12, pp. 14–28.
[114] J. YANOSIK AND T. MCCRACKEN, *A nine-point, finite difference reservoir simulator for realistic prediction of adverse mobility ratio displacements*, Soc. Pet. Eng. J., 19 (1979), pp. 253–262.
[115] W. F. YELLIG AND L. E. BAKER, *Factors affecting miscible flooding dispersion coefficients*, J. Canad. Pet. Tech., 20 (1981), No. 4, pp. 69–75.
[116] L. C. YOUNG, *A finite-element method for reservoir simulation*, Soc. Pet. Eng. J., 21 (1981), pp. 115–128.

CHAPTER III

A Front Tracking Reservoir Simulator, Five-Spot Validation Studies and the Water Coning Problem

JAMES GLIMM, BRENT LINDQUIST, OLIVER McBRYAN, AND L PADMANABHAN

1. Introduction. The goal of front tracking, for problems containing important discontinuous behavior, is to combine the flexibility of a general purpose hydrodynamics calculation with the efficiency of a special purpose calculation. This goal is currently being pursued by the authors in problems of oil reservoirs, gas dynamics and fluid interface instabilities.

Reservoir simulation problems involve the numerical fluid dynamics of multiphase flow, specifically the computation of the flow of oil, water and/or gas in a porous medium (sandstone). The fluid variables for these problems are pressure and the phase concentrations (saturations). The phases are oil, water and gas. Each rock pore is thought of as containing a single phase, while a macroscopic region (containing many pores) is characterized by its fractional volume contents of the three phases (i.e. by the saturations of the three phases). Typical "shocks" in such problems are discontinuities in phase concentrations: the oil, gas or water banks. The water coning problem concerns the flow field in the neighborhood of a production well in an oil reservoir. It is relevant in the primary recovery stage of reservoir production, during which oil production arises solely from the internal reservoir pressure. Initially the oil floats above a layer of water. As oil is extracted from the well, the water table in the vicinity of the well rises, and eventually reaches the inlet of the well. Subsequently, the well produces water in addition to (in the worst case, instead of) oil. The phenomenon is to some extent weakened by gravity, which acts against the pressure gradient pulling the water upwards towards the well outlet. The coning phenomenon may also be affected by capillarity, which adds a nonlinear, diffusion-type term to the reservoir equations. We ignore the effects of capillarity in this paper.

Standard solutions to this problem are obtained by finite difference techniques based on iterative inversion methods for coupled elliptic–hyperbolic systems. The methods proposed here differ in certain respects and are designed with the purpose of eliminating certain difficulties of these standard methods. Our basic strategy is to introduce an interface as a computational degree of freedom in the calculation. The basic variables are allowed either to be discontinuous (oil, water saturation) or have discontinuous derivatives (pressure) across these interfaces.

Due to their nonlinear dependence on pressure and saturation, the defining equations also have discontinuous behavior across an interface. We use the interface as an intrinsic part of the solution algorithms for both the pressure and saturation equations. This strategy is termed "front tracking."

This front tracking method has been used to study horizontal reservoir flow problems [1], [2], [3], [11]. The attraction of this method is the resolution of fingering instabilities it provides while using a relatively coarse computational grid. The water coning problem is representative of problems which arise in contemporary reservoir engineering practice. In this paper we describe the application of the front tracking scheme to this problem and show that it describes the coning accurately even with relatively course grids. This is part of a series of increasingly difficult problems to which the tracking method should make a contribution. Of greater difficulty are the gas coning problem, which introduces compressibility effects, and the percolation problem, which follows a gas bubble rising within the reservoir. Tracking is also being used currently to study the Rayleigh–Taylor and Kelvin–Helmholtz instabilities in the nonlinear regime [12], [13], and to study shocks and contact discontinuities in gas dynamics [14]. Although tracking has been used before [4] (see also [5] for a related method using tracked characteristics), the present effort is, to the authors' knowledge, unique as a development of front tracking into a general purpose hydrodynamics code, applicable to a range of problems.

In §2 we present the equations used to model the reservoir flow field in the cylindrical geometry appropriate for the coning problem. We discuss the front tracking scheme and the method of solution of the resulting equations. As much of the front tracking scheme has been previously reported [2], the discussion will be brief. In §3 we discuss the improvements in the front tracking scheme that have been introduced recently. In §4 we show the results of the quarter five-spot validation studies performed to test the improved scheme. Section 5 describes the results of two test problems run under simplified coning conditions. The results for the coning problem proper are shown in §6. Our conclusions appear in §7.

2. Problem formulation and solution methods.

2.1. The equations.
If the oil and water are treated as incompressible, the two equations governing the reservoir coning problem describe the conservation of fractional water volume, s, (the fractional oil volume being $1 - s$) and the total fluid velocity flow field, $\mathbf{v}$,

(2.1a) $$\frac{\partial s}{\partial t} + \nabla \cdot \mathbf{F}(s, P) = \text{sources},$$

(2.1b) $$\nabla \cdot \mathbf{v}(s, P) = \text{sources}$$

where P is the reservoir pressure field. The fractional flow function

$$(2.2) \qquad \mathbf{F} = f(s)\mathbf{v} - \frac{\lambda_w \lambda_o}{(\lambda_w + \lambda_o)} (\rho_w - \rho_o) g \mathbf{K} \cdot \hat{z}$$

consists of a gravity free transport term $f(s)\mathbf{v}$ and a gravitational flow modifying term. The total fluid velocity is given by Darcy's law, which expresses fluid velocities as proportional to reservoir pressure and gravitational potential gradients. Neglecting the effects of capillary pressure, the total fluid velocity is given by

$$(2.3) \qquad \mathbf{v} = -(\lambda_w + \lambda_o)\mathbf{K} \cdot \nabla P - (\lambda_w \rho_w + \lambda_o \rho_o) g \mathbf{K} \cdot \hat{z}.$$

In (2.2) and (2.3), the rock permeability tensor $\mathbf{K}(\mathbf{r})$ describes the rock properties of the reservoir including spatial variations, the most notable of which are the fairly rapid variations in the vertical direction due to geological layers. Here $\lambda_i = k_i(s)/v_i$, v_i, $k_i(s)$ and ρ_i denote respectively the transmissibility, viscosity, relative permeability function and phase density of fluid phase i, while g is the gravitational acceleration constant and $\hat{z}$ is a unit vector in the vertical (upwards) direction. The gravity free water fractional flow function $f(s)$ plays a determining role in modeling the reservoir, and requires some discussion. In the derivation of (2.1), (2.2) and (2.3) from basic physical laws (plus Darcy's law), $f(s)$ is determined to have the form

$$(2.4) \qquad f(s) = \frac{\lambda_w(s)}{\lambda_w(s) + \lambda_o(s)}.$$

The λ_i are, however, unrestricted; the functional forms chosen for the λ_i depend on the type of reservoir being simulated. Reservoir rock matrix consists of multiple interconnected pores. In this complicated topology, pore channels occupied by water reduce the ability of oil to flow, and vice versa. A simple, qualitatively correct model for an oil-water reservoir is

$$(2.5) \qquad k_w(s) = s^2, \qquad k_o(s) = (1 - s)^2.$$

The nonlinearities in (2.5) model this interference phenomenon, generally referred to as immiscible flow. Substitution in (2.4) gives

$$(2.6) \qquad f(s) = \frac{s^2}{v_w \lambda}, \qquad \lambda = \frac{s^2}{v_w} + \frac{(1-s)^2}{v_o}, \qquad \text{immiscible flow.}$$

In the simulation of a real reservoir, experiments on rock samples taken from the reservoir in question are used to determine the functional forms in (2.5).

The case of miscible displacement also occurs, with a solvent replacing water as the displacing fluid (CO_2 tertiary recovery is an example of miscible displacement in current practice). In this case the solvent and oil do not interfere with each other's flow and (2.5) becomes

(2.7) $$k_w(s) = s, \qquad k_o(s) = (1 - s).$$

In such a case each pore may be occupied by a mixture of the two phases, the pure phase viscosities v_i are no longer relevant, and a mixture viscosity $v(s)$ is introduced,

(2.8) $$v(s) = (v_w^{-1/4} s + v_o^{-1/4} (1 - s))^{-4}.$$

This formula is phenomenological, the exponents depending on assumptions about the degree of microscopic mixing. Interpreting $\lambda_i(s)$ as $k_i(s)/v(s)$ in the miscible displacement case, (2.8) and (2.7) yield

(2.9) $$f(s) = s, \qquad \lambda = v(s)^{-1}, \qquad \text{miscible flow.}$$

The form of (2.8) will play little role in our miscible displacement calculations as, due to our choice of initial conditions and the lack of diffusion, s takes on the values 0 and 1 only.

The source/sink terms in (2.1) are the injection and production wells. In the coning problem we have replaced the sources and sinks by an appropriate choice of boundary conditions; the interior of the computational region is therefore source/sink free. At the $r = r_{\max}$ boundary of the computational region we model inflow from infinity into the drainage basin of the producing well under study by imposing Dirichlet conditions in the pressure equation (2.1b). The value of this imposed boundary pressure must ultimately come from some other study or field data; we approximate by imposing a constant (far field) pressure value along the full side. The bore of the production well under study is located at $r = r_{\min}$. Here we impose zero Neumann data (no flow) over most of the side except for short Dirichlet segments to model the well inlets. Over these short Dirichlet segments the pressure is held constant as representing an average internal well pressure. These well boundary conditions are an idealization to common engineering data wherein a single well head pressure (or flow rate) can be specified, but not the pressure contour down the well shaft.

2.2. The coordinate system. Typical reservoirs are not homogeneous. The most prominent heterogeneities are the distinct geological layers. The rock properties can vary between layers by factors of 10 to 100 or more. Within a layer, the rock properties will in general also vary, but usually in a less dramatic fashion. Further complications are that the tangent plane to a layer (the bedding plane) is not in general horizontal, nor, in fact, need the well be vertical. In particular the well is not in general normal to the bedding plane.

As a simplifying assumption, we take the well to be vertical, the bedding plane horizontal. The coning problem is then rotationally invariant and is best solved in cylindrical coordinates. The equations become two-dimensional, expressed in r

and z coordinates. As a further simplification (see the companion paper [11] in which this simplification is dropped), we take the reservoir to be homogeneous separately in the r and z directions. Then the rock permeability tensor $\mathbf{K}$ becomes diagonal.

In cylindrical coordinates the pressure equation (2.1b) is

$$(2.10) \qquad -\frac{1}{r}\frac{\partial}{\partial r} r \Lambda_{rr} \frac{\partial P}{\partial r} - \frac{\partial}{\partial z} \Lambda_{zz} \frac{\partial P}{\partial z} = \frac{\partial}{\partial z}(\lambda_w \rho_w + \lambda_o \rho_o) g K_{zz},$$

where $\Lambda \equiv (\lambda_w + \lambda_o) \mathbf{K}$. The factors $r^{\pm 1}$ result from expressing the divergence in the non-Euclidean metric $dr\, d\theta$. The substitution

$$(2.11) \qquad x \equiv r^2, \quad y \equiv z, \quad \Lambda_{xx} \equiv 4r^2 \Lambda_{rr}, \quad \Lambda_{yy} \equiv \Lambda_{zz}$$

restores the Euclidean metric and transforms (2.10) to

$$(2.12) \qquad \frac{\partial}{\partial x} \Lambda_{xx} \frac{\partial P}{\partial x} + \frac{\partial}{\partial y} \Lambda_{yy} \frac{\partial P}{\partial y} = \frac{\partial}{\partial y}(\lambda_w \rho_w + \lambda_o \rho_o) g K_{yy},$$

which has a symmetric leading term and vanishing first order derivatives in P. Note that a jump discontinuity in s (and hence in $\lambda(s)$) leads to a jump in the RHS of (2.10) and (2.12) and hence to a "line source delta function" as one passes across the saturation discontinuity in a direction normal to the front.

2.3. Front tracking. There are three essential features of our front tracking method. In the elliptic pressure equation, the use of mesh alignment keeps the discontinuity interface on mesh lines or mesh diagonals, thus ensuring that it lies along edges of the finite element triangulation (see below for an expanded discussion on this point). In the hyperbolic conservation equation for the water saturation, the tracking method is equivalent to the adoption of local normal and tangential coordinates. The use of Riemann problem solutions in the normal direction guarantees the correct resolution of discontinuities into coherent waves. (This is in effect a nonlinear normal mode expansion.) The third feature of our method addresses the programming complexity issue. Use of data structures with supporting subroutine libraries allows convenient extension, modification and maintenance of front tracking algorithms. Modern structured programming and computer science concepts also lead to computationally efficient front tracking support algorithms. In fact, with computation dominated by the solution of large linear systems of equations, the overhead support for tracking is negligible.

Through an IMPES (implicit pressure, explicit saturation) scheme, the coupled pressure and saturation equations are solved separately (sequentially) in each timestep. The justification for the sequential solution of the coupled system of reservoir equations is that the saturation and pressure variables describe changes which occur on sharply different time scales. Pressure transients disappear in the reservoir fairly quickly. After opening a well, pressure equilibrium will be reached in the order of a day, while saturation changes occur over the time scale of the life of the reservoir (some number of years). Assuming the fluids are incompressible implies that the pressure reaches equilibrium instantaneously, relative to a given saturation profile. In regions of rapid flow, such as near the well opening, these two time scales can become comparable and sequential solution methods may be inappropriate.

In order to discretize the pressure equation, the two-dimensional computational grid is dynamically aligned to the moving saturation discontinuity interface, so that the discontinuity lies exactly on grid lines of the pressure equation. The construction of the mesh-aligned grid is an essential feature of our method [6]. We outline the main steps in its construction. First a rectangular mesh is chosen. It need not have uniform grid spacing, so that separate mesh refinements in the two coordinate dimensions can be introduced at this stage. The interface is defined as a set of points joined by linear (or polynomial) line segments. Each intersection point between a rectangular grid line and an interface line segment (a bond) is determined, and inserted as a new point on the interface. Going through this augmented interface in bond sequence, one of two actions is taken at each vertex (bond endpoint). Either a node of the rectangular grid is displaced, so as to coincide with the interface vertex, or the interface vertex is deleted, by substitution of a single bond for two consecutive bonds. Various priority functions control this choice and other aspects of the grid construction. A final check on aspect ratios of the resulting triangles is then performed to ensure that the construction is reliable. In our experience, failure to align the pressure equation grid to these discontinuities results in inaccurate flow velocities. The importance of this effect depends on the size of the mobility ratio across the discontinuity. If the mobility ratio is greater than 2, alignment appears to be essential. Though the resulting aligned grid is a grid of irregular quadrilaterals, a rectangular index structure is preserved. In addition, the grid is actually rectangular away from the interface, irregular quadrilaterals occurring only along the front. As a result of these two facts, fast iterative solution methods for the pressure equations can be used effectively.

The pressure equation is discretized using finite elements. The irregular (nonrectangular) quadrilaterals aligned along the discontinuity interface are bisected into two triangles. If the interface passes through a diagonal of the quadrilateral, then this is the diagonal chosen for bisection. Otherwise, the shortest diagonal or (for nonconvex quadrilaterals) the interior diagonal is chosen. For regular (rectangular) quadrilaterals, there is no bisection. Thus if linear (quadratic, cubic) elements are used on the triangulated region, bilinear (biquadratic, bicubic) elements can be consistently used on the rectangles. Multi-grid, accelerated conjugate gradient and direct solution methods have all been used to solve the discretized equations. The choice of method and acceleration operator is problem and grid size dependent and will not be discussed here; see [15].

For the hyperbolic equation there are three issues: the solution in regions away from the front (the interior region problem), the propagation of the interface points, and the coupling of the interior scheme to the front (equivalently the reflection of waves by, and their transmission through, the front). The interior problem is standard and can be solved by, for example, operator-split upwind, Lax–Wendroff or random choice methods. (In the case of cylindrical coordinates, the space operator splitting is done in the r, z coordinates.) For the quarter five-spot runs discussed in §4, the interior problem was solved with the operator split random choice scheme. The coning runs described in §6 used an operator split first order upwind scheme for the interior problem. The coupling of the interior to the front is discussed in [2]. The propagation of the front uses Riemann problem solutions based on data taken from both sides of the front.

2.4. The Riemann problem and wave speeds. The (analytical) solution to the Riemann problem for a Buckley–Leverett equation in the presence of gravity has been discussed in [7] for the case of one space dimension. This 1-D solution is used to propagate the discontinuity in its normal direction. As we are dealing with incompressible media, the hyperbolic problem (2.1a) keeps a fairly simplified form. Using (2.2), we can write (2.1a) as

$$(2.13) \qquad \frac{\partial s}{\partial t} + \nabla \cdot (f(s)\mathbf{v} - g(s)\hat{z}) = 0,$$

where the notation $g(s)$ is clear from (2.2). As we model a source-free region in the coning problem, the velocity field is divergence-free and (2.13) can be written

$$(2.14) \qquad \frac{\partial s}{\partial t} + v_r \frac{\partial f(s)}{\partial r} + v_z \frac{\partial f(s)}{\partial z} - \frac{\partial g(s)}{\partial z} = 0.$$

The simplification evident in (2.14) is that for incompressible flow, no pseudo-source term arises from the cylindrical geometry. The divergence operator in (2.14) can be treated as a cartesian operator in the r, z coordinates and the velocity field can be treated as a local constant in the solution of the Riemann problem.

3. New aspects of the front tracking scheme. Several new additions to the front tracking scheme have been implemented either as general improvements or because of the particular nature of the coning problem. We discuss them here.

3.1. Higher order elements in the pressure solution. Use of higher order finite elements (see [16]) for the pressure equation solution accomplishes two purposes. Except for physical discontinuities at the interface, higher order elements ensure that the velocity field is continuous. Their use also improves the resolution of the pressure and velocity singularity in a neighborhood of an outlet at the production well. At the producing well opening, the singularity is approximately

(3.1) $\quad\quad P \approx ((r^2 + z - z_0)^2)^{-1/2}, \quad\quad v \approx (r^2 + (z - z_0)^2)^{-1}.$

For the coning problem, it was found that quadratic/biquadratic elements gave much more satisfactory results for the pressure (and therefore velocity) fields near the well singularity than could be obtained using linear/bilinear elements.

In the five-spot runs reported here, the well singularity did not play a large role in our calculations since the runs were begun with the interface away from the injection region and stopped when the interface reached the producing well (breakthrough). The continuation past breakthrough is of practical engineering interest but is not required for our validation purposes. To obtain satisfactory answers near the interface, use of linear/bilinear elements combined with tangential velocity regularization (see §3.2) is sufficient and appears to be more efficient than the use of higher order elements. However, the higher order elements are more robust and also provide a check on the averaging used with the linear/bilinear elements.

In the interior, away from the well opening and the interface, there is little variation in the solution, and any of the finite elements considered above are sufficient.

To be fully effective, the higher order elements should be based on curvilinear quadratic or cubic) triangles, to eliminate unphysical singularities in the pressure P and velocity v at vertices between bonds of piecewise linear interfaces. This aspect of higher order elements has been implemented [15] but is not fully tested and was not used in the runs reported here.

3.2. Tangential regularization of interface velocities. The propagation of the interface requires the use of a velocity $v = -\lambda \nabla P$ which, for reasons of stability, should be separately continuous on each side of the discontinuity. We discuss two cases: pressure computed with linear/bilinear elements and pressure computed with quadratic/biquadratic elements. In the former case, $\lambda \nabla P$ is piecewise constant, and is not smooth enough for propagation of the interface in unstable cases (mobility > 1). We have employed a tangential average with a three or five point stencil (depending on grid size). In Fig. 1 we show an example of raw (piecewise constant) and averaged velocities. Here $|v|$ is plotted vs. arclength for a typical time step in a miscible displacement, $M = 5$, diagonal five-spot run on a 32×32 grid. In this case a three-point averaging stencil was used. Because of local mesh refinement, the averaging stencil extends over a sufficiently short distance and the peak in $|v|$ along the diagonal of the computational square is preserved by the average.

If quadratic/biquadratic elements are used for the pressure solution, $v = -\lambda \nabla P$ is piecewise linear. However (as presently implemented), these higher order elements are still based on linear bonds, which results in velocity singulari-

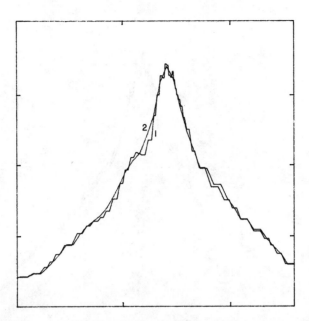

FIG. 1. *Velocity magnitude vs. parametric position along the discontinuity interface for a time step approximately ⅔ to breakthrough in a diagonal five-spot run for a 32 × 32 grid. Graph 1 is of the "raw" velocity as returned directly from the elliptic solution for P using linear/bilinear elements. Graph 2 shows the effect of a three-point stencil tangential averaging.*

ties at the vertices of the linear bonds in the exact (continuum) solution of the pressure equation, and in jump discontinuities in ∇P across bond vertices. To overcome this problem, we evaluate ∇P only at bond midpoints and, using a three point stencil, construct a continuous $\lambda \nabla P$ using averaging. In Fig. 2 we compare bond midpoint raw velocity (before tangential smoothing) solutions along the interface using linear/bilinear elements and using quadratic/biquadratic elements. The run is an $M = 5$, miscible displacement, diagonal five-spot run on a 32 × 32 elliptic grid (the linear/bilinear plot is not from the the same run as produced Fig. 1, but is almost an identical run). The quadratic/biquadratic derived velocity profile is much smoother than the linear/bilinear. At the velocity peak, however, the quadratic/biquadratic velocity profile becomes a bit jagged, indicating that some tangential smoothing is required in the finger tip region even in the higher order element case.

3.3. 1-D mesh refinement. Independent one-dimensional mesh refinement in each of the two coordinate directions is an option in the elliptic grid construction. This feature is quite important in the coning problem for resolving the singularity near the well opening. In fact, we have found mesh refinement alone sufficient to

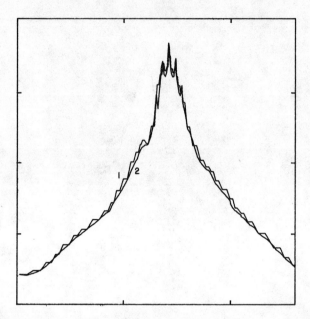

FIG. 2. *Velocity magnitude vs. parametric position along the discontinuity interface for a time step approximately ⅔ to breakthrough in a diagonal five-spot run for a 32 × 32 grid. Graph 1 is of the "raw" velocity as returned directly from the elliptic solution for P using linear/bilinear elements. Graph 2 is of the "raw" velocity as returned directly from the elliptic solution for P using quadratic/biquadratic elements.*

reproduce the $\log(r)$ and r^{-1} pressure singularities in our test studies. In this way, we have been able to avoid adding problem-dependent elements to the finite element space used to solve the pressure equation (see §5).

The mesh refinement algorithm has further uses. It can provide simple concentration of grid lines along the discontinuity interface in order to improve the pressure (and hence velocity) solution across the discontinuity. This is especially valuable in cases where the mobility ratio across the discontinuity is large. It can be used to counter the effects of choice of computational coordinates. For example, the use of the computational coordinate $x = r^2$ would imply that equally spaced x intervals would correspond to r intervals refined near $r_{\max}$, i.e. refinement away from the production well region! In the studies reported here, the mesh refinements are tailored to achieve some combination of these objectives.

4. Validation studies–the quarter five-spot problem. A standard reservoir test problem is the quarter five-spot problem: a square domain in the horizontal plane, with injection and production wells at opposite corners along one of the diagonals. For this problem we demonstrate convergence of our method under mesh refinement and invariance under change of mesh orientation. Conventional reservoir simulators have difficulty with these tests especially for mobilities $\gg 1$ [8], [9], [10]. For the special case of miscible displacement, mobility ratio 1, the hyperbolic and elliptic equations (2.1) decouple and can be solved in closed form. In these cases we obtain agreement with the exact solution, even on rather coarse grids.

Perhaps the most useful geometry that has been found for solving the quarter five-spot problem involves the orthogonal coordinate system (ψ, η), where $\psi(x, y)$ = const. are the velocity equipotentials and $\eta(x, y)$ = const. are the flow lines in the miscible displacement, unit mobility ratio case [9]. As the conformal mapping from cartesian coordinates to this curved geometry involves elliptic integrals, we shall refer to this geometry as the "elliptic" coordinate system. This geometry is also useful for miscible flow, mobility ratio $\neq 1$; in our case we shall use a mobility ratio of 5, a fairly unstable flow. Figure 3 shows the tracked oil-water discontinuity front at selected time steps for three mesh sizes for runs using this "elliptic" coordinate system. Linear/bilinear elements were used for

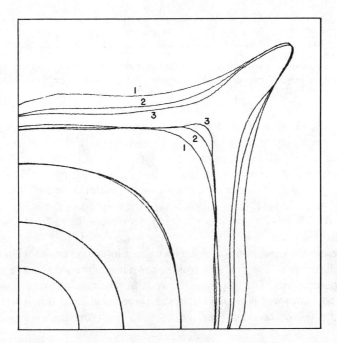

FIG. 3. *Front position at selected times for runs with three mesh sizes in the velocity equipotential ψ, streamline η coordinate system simulation of a quarter five-spot reservoir. The mesh sizes are 1–16 × 16 grid, 2–32 × 32 grid, 3–48 × 48 grid. These mesh sizes refer to the grid used to solve equation* (2.1b). *The grids involved in the solution of* (2.1a) *were, respectively,* 12 × 12, 24 × 24 *and* 36 × 36.

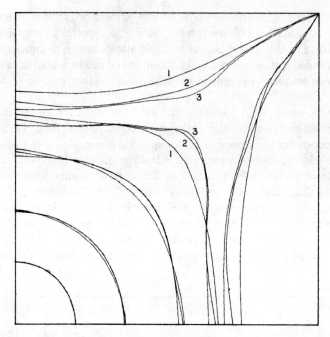

FIG. 4. *Front position at selected times for runs with three mesh sizes in the quarter five-spot reservoir modeled using a cartesian coordinate system. The runs correspond to the same mesh sizes as in Fig. 1.*

the elliptic solution in each case while a three-point stencil was used for the tangential averaging of front velocities. The fronts marked "1" correspond to a run having a 16 × 16 mesh refined grid for the elliptic equation (2.1b). The fronts marked 2 correspond to a run with a 32 × 32 elliptic grid. The fronts marked 3 correspond to a run with a 48 × 48 elliptic grid. In each case the grid used to solve the hyperbolic equation was ¾ the fineness of the elliptic grid (i.e. for run 1 the hyperbolic grid was 12 × 12). However, due to front tracking the hyperbolic grid plays little role in the front propagation. The number of points used to define the front position in each run had inter-point distances consistent with the elliptic grid. The first four fronts shown in each run are for the same four equally spaced times. The last front shown is for the breakthrough time in each case.[1] These times will not coincide in the three runs. That this is so of course comes from the observation that the 48 × 48 grid run has a more narrowly

---

[1] The pressure solution becomes infinite at the point source wells. For the "elliptic" coordinate system this problem is solved by cutting a small region out from around a well using one of the equipotential lines near the well and replacing the point source well by constant pressure (Dirichlet) conditions on this cut-off contour. Breakthrough is then defined as the time at which the front first hits the production well cut-off contour. For clarity, these cut-off contours are not displayed in Fig. 3.

defined finger than the coarser grid runs and leads the other two. The agreement between runs 2 and 3 is very good; the coarseness of grid 1 only begins to make itself felt near the end of the simulation.

Figure 4 shows the same runs as above, but now in the cartesian x, y-coordinate system. Whereas in the "elliptic" coordinate system the front moves essentially perpendicular to the ψ-coordinate axis, in this system the fingered region is moving roughly at 45° to the axes. This is known as the diagonal five-spot geometry. The agreement between the runs at different mesh sizes is as good in this geometry as for the "elliptic" coordinate system.

Figure 5 plots the 32 × 32 elliptic grid runs from Figs. 3 and 4 together for ease of comparison. Since the run of Fig. 4 is driven by a constant flow rate (δ function source and sink of determined strengths) whereas that of Fig. 3 is driven by a constant pressure drop between the cut-off circles around the wells, the computational times in the two runs do not (necessarily) match. As a simple approximation we postulate that the difference in treatment of source terms mainly affects the strength of the source, and can be compensated for by a rescaling of time. Thus, a time scaling parameter α is introduced, such that $t_1 = \alpha t_2$ where t_1 and t_2 are the times in the two runs. For comparison between the two runs, α is chosen to provide agreement in the early time steps, and held fixed

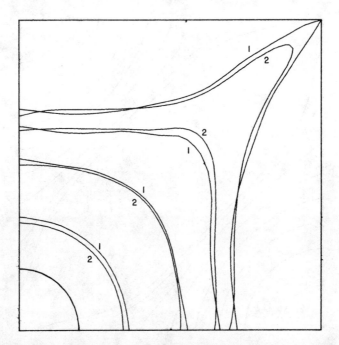

FIG. 5. *Front position at selected time steps for the runs numbered 2 in FIGS. 3 and 4, superimposed to show the effects of change of coordinate system.*

throughout the comparison. (An examination of Fig. 5 suggests that even after such a constant time rescaling, the difference in boundary conditions could still be the major cause of difference between the runs.) Of greater importance however is the agreement of the front shapes in the two geometries, maintained throughout the simulation. The "elliptic" coordinate system produces a slightly sharper finger which moves slightly faster than in the cartesian system. There appears to be an unresolved problem of a slight boundary drag of the front in the "elliptic" system simulation.

The final study we run is for a reservoir consisting of a square domain with injection wells at the two corners along one diagonal and production wells at the two corners along the other diagonal. This is referred to as parallel five-spot geometry. Here the coordinate system is again cartesian but the direction of front movement with respect to the grid axes is different than in the diagonal five-spot geometry. Fig. 6 shows the results for a run in this geometry. Mesh refinement in this geometry produces results comparable with Figs. 3 and 4 and we do not reproduce them here. Except for the geometry all parameters are as for the runs marked "2" in Figs. 3 and 4. Fig. 7 superimposes those fronts expanding from one of the injection wells in Fig. 6 onto run "2" of Fig. 3. Fig. 8 shows the same superposition, now for run "2" of Fig. 4. (Note, front positions at twice as many

FIG. 6. *Front position at equidistant time steps for a run in the parallel five-spot geometry referred to in the text. The run parameters are the same as for the runs numbered 2 in Figs. 3 and 4.*

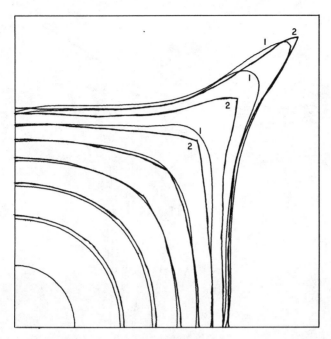

FIG. 7. *Superposition of Fig. 6 and the runs corresponding to run 2 of Fig. 3 showing the effects of change of coordinate system. The fronts indicated by 1 correspond to Fig. 3, those indicated by 2 correspond to Fig. 6.*

time points are plotted in Figs. 6 through 8. Every second time step plot had been deleted in Figs. 3 and 4 for clarity.) The general agreement between the fronts in the three geometries is very good. Slight differences in widths of the central finger are noticed, with the diagonal geometry producing the widest finger and the "elliptic" geometry the narrowest.

5. Validation studies–the coning problem.

5.1. A one-dimensional study. As stated in §2.1, the production well in the coning problem is modeled by boundary conditions at $r = r_{min}$. No flow in the radial direction (zero Neumann boundary conditions) is specified along most of the $r = r_{min}$ boundary except for one or more short Dirichlet segments intended to model the actual well outlet(s). Here a constant pressure is specified, the pressure being some averaged well pressure. The $r = r_{max}$ boundary has a constant pressure specified (Dirichlet boundary conditions), modeling far field conditions. The top and bottom boundary conditions specify no flow through the boundary (zero Neumann conditions). If the mixed $r = r_{min}$ boundary conditions are replaced by a single Dirichlet condition of constant pressure, then for a homogeneous reservoir where gravity is ignored, and given an initial oil–water discontinuity front which is a vertical straight line with constant saturation

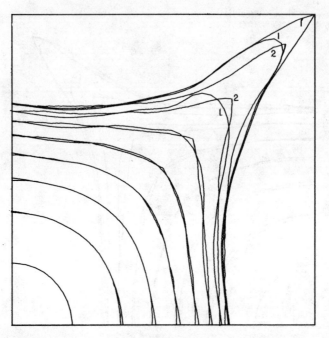

FIG. 8. *Superposition of Fig. 6 and the runs corresponding to run 2 of Fig. 4 showing the effects of change of coordinate system. The fronts indicated by* 1 *correspond to Fig. 4, those indicated by* 2 *correspond to Fig. 6.*

values on either side, the problem is one-dimensional. If miscible displacement (2.9) is used, equation (2.1b) can be solved exactly for arbitrary mobility ratio across the discontinuity. The expression is messy and contains nothing of interest that is not given by simplifying to unit mobility ratio. Using unit mobility ratio the solution to (2.1b) is trivially calculated as

$$(5.1) \qquad P(r, z) = P(r) = P_a \log\left(\frac{r}{r_{max}}\right) + P(r_{max})$$

where P_a is simply given in terms of the boundary radius values and pressures. The logarithmic divergence in (5.1) provides two problems. First, trivially, one cannot model to $r = 0$ but only to a small cut-off value. The second problem is more critical. Using a finite element solution of (2.1b) employing polynomial elements of degree two, one cannot hope to do an adequate job of representing (5.1). We found it easiest to surmount this problem by using our mesh refinement facility. The mesh in the r direction was given a logarithmic weighting of the form

$$(5.2) \qquad \text{prob}(r) \approx \log\left(\frac{r}{r_{min}}\right)$$

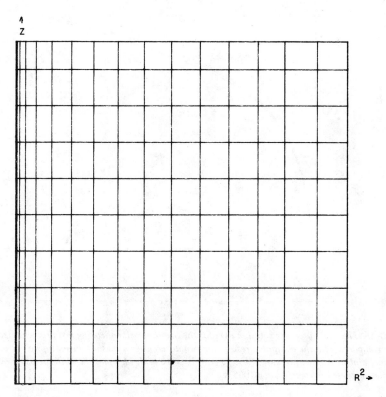

FIG. 9. *An example of a one-dimensionally refined mesh used in the one-dimensional coning test problem discussed in §5. The grid is 20 × 10. A logarithmic density function is combined (with equal weight) with a regular grid (constant density function) to give the resultant grid in the r direction. The grid in the z direction is regular.*

where prob(r) is the probability of a mesh line occurring between r_{min} and r. This mesh refinement was combined (with equal weighting) with a regular mesh. Fig. 9 displays a typical grid obtained for a 20 × 10 mesh. (There are 19 internal grid lines in the r direction.) Fig. 10 compares the exact pressure result (5.1) with the results computed from two elliptic grid mesh sizes, 20 × 10 and 40 × 10. The r_{min} boundary pressure is taken to be 1400 psia, the r_{max} is 1800 psia. The radial extent of the reservoir is $r_{min} = 5$ feet, $r_{max} = 1005$ feet. Fig. 11 compares the exact velocity $v(r) = -\text{const.} \, \partial P/\partial r$ with the results computed from the same two grids in Fig. 10. In the above formula for $v(r)$ the constant is given by the rock permeability divided by a reference viscosity, in this case that of water.

As expected, the quality of the solution deteriorates tremendously if the logarithmic density weighting is not used in the r direction for the solution to (2.1b).

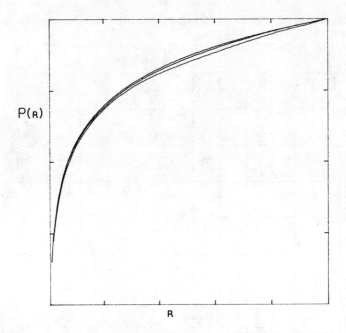

FIG. 10. *The exact pressure solution $P(r)$ for the one-dimensional coning test problem discussed in §5 compared with the computed results for two different grid meshes. From top to bottom, the pressure plots are: exact pressure, 40×10 grid, 20×10 grid. This plot shows the pressure over the complete modeled range $5 \text{ ft} \leq r \leq 1005 \text{ ft}$.*

5.2. An approximately spherically symmetric study. We now examine a study where the producing well boundary condition is replaced by no-flow[2] boundary conditions except for a very small Dirichlet segment of height $2d$ in the center of this side. The top and bottom boundary conditions are replaced by Dirichlet (constant pressure) boundary conditions, the same pressure value being imposed as is used on the far field boundary. The initial water–oil discontinuity is a half-circle of radius R centered on the short Dirichlet segment. (The Dirichlet segment on the well boundary is to be thought of as a small half circle of radius d centered on the middle of the boundary.) Then, for a homogeneous reservoir in the absence of gravity, the solution in the infinite mobility ratio limit is

$$(5.3) \qquad P(r) = \begin{cases} \dfrac{P_o}{r} + P(d), & d \leq r \leq R, \\ P(\text{far field}), & R \leq r, \end{cases}$$

where P_o is easily expressed in terms of $P(d)$, $P(\text{far field})$, R, and d. The radius r in (5.3) is, of course, calculated with respect to the center of the well outlet (as

---

[2] As always we mean to flow perpendicular to the boundary.

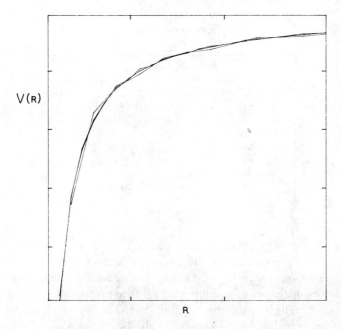

FIG. 11. *The exact velocity solution $v(r)$ for the one-dimensional coning test problem discussed in §5 compared with the computed results for the same two runs shown in Fig. 10. This plot shows a blow up of the region 5 ft $\leq r \leq$ 155 ft. The exact and 40 × 10 solutions are almost indistinguishable. Outside the range of r shown all three plots become almost indistinguishable.*

opposed to the center of the short Dirichlet segment). The velocity in each region is radially directed and, using Darcy's law, follows directly from (5.3).

Because the pressure diverges as r^{-1} and the velocity as r^{-2} at the centered outlet, the use of a short Dirichlet segment to represent a small half circle implies that the simulation can only approach the exact answer (5.3) near the well in the limit that the Dirichlet segment goes to a point. For r sufficiently larger than d however, (5.3) does represent an "exact" solution to the modeled problem. Within the limits of the proviso stated above, our results in this study show the same convergence properties to the "exact" solution (5.3) as we showed in §5.1. Thus we do not include plots of the results here.

6. The water coning problem. We present the results for a series of three runs for a coning simulation. Immiscible flow conditions (2.6) were used throughout; the (constant) reservoir parameters used in the simulations are given in Table 1.

The first run was a calculation in the cylinder 10 ft $\leq r \leq$ 200 ft, $z_{\text{low}} \leq z \leq z_{\text{low}} + 50$ ft centered about a production well bore. (We scale the z-axis of the computation such that $z_{\text{low}} = 0$.) The effective well opening at the $r = 10$ ft boundary was taken to be 5 feet wide, lying in the region $35 \leq z \leq 40$.

TABLE 1
*The reservoir parameters used
in the coning simulations.*

Reservoir parameters	
density	
oil	50 lbs/ft^3
water	65 lbs/ft^3
viscosity	
oil	5 cp
water	1 cp
rock permeability	
r direction	10 mD
z direction	1 mD

The pressure across this effective outlet was held constant at 1500 psia throughout the run. The pressure at $r = 200$ boundary of the modeled region was held constant at 2200 psia throughout the calculation. Calculation has shown that this is not an unreasonable value assuming the far field pressure (i.e. at say 1000 feet from the outlet well) is of the order of 3000 psia. The hyperbolic equation (2.1a) was solved on a regular, 48×48 grid. The elliptic equation (2.1b) was solved on a 32×32 grid which was differentially refined in the r and z directions. Figure 12 shows a plot of the elliptic grid at a time in the simulation near breakthrough, demonstrating the differential mesh refinement in the two coordinate directions. The r direction mesh refinement was defined using a weighting of grid lines which was a combination of $\log(r/200)$ and constant density functions. The z direction mesh refinement used a combination of $1/|z - 40|$, constant and front weighted density functions. The front-weighted density function ensures additional grid lines around the interface for adequate discontinuity resolution. Note that the abcissa coordinate in this plot is the square of the radius, thus the cone shown in Fig. 12 appears narrower than it actually is.

Figure 13 shows the developing cone at intervals of 926 days except for the final plot cone which is shown at breakthrough (2,550 days). Noteworthy is the appearance of the oil counterflow in the region $135 \leq r$, where the gravitational gradient is stronger than the pressure gradient. The no-flow conditions at the lower boundary restrict the downward advance of this countercurrent. The measured half-angle of the cone for the breakthrough contour is approximately 30°. Figure 14 is a plot of the breakthrough pressure field. The outlet region $35 \leq z \leq 40$ at the $r = 10$ boundary is very sharply delineated by the pressure solution. For further clarity, Figure 15 replots the pressure at breakthrough by showing the contours of constant $P(r, z)$. Figure 16 is a plot of the saturation $s(r, z)$ at breakthrough. The oil rarefaction in the counterflow region can be seen as well as the rarefaction in the water cone. (Note that the saturation here has been

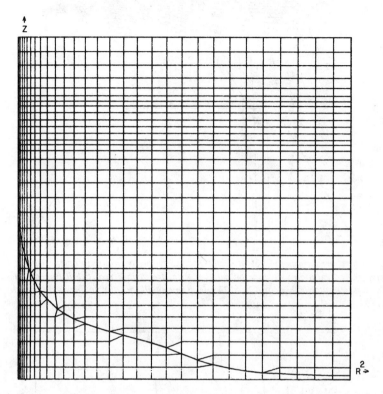

FIG. 12. *An example of the elliptic mesh generation for a time point in the* $10 \leq r \leq 200$ *reservoir simulation discussed in the text. The mesh is* 32×32. *The contributions of the various weighting densities* $(1/|z-z_{\text{outlet mid pt}}|$, *constant and front-weighted in the z direction;* $\log(r/r_{\max})$ *and constant in the r direction) used to generate the mesh can be distinguished in the plot. The front-weighted density function ensures additional grid lines around the interface for adequate discontinuity resolution.*

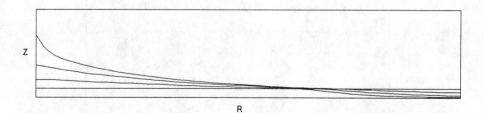

FIG 13. *The developing cone shown at intervals of 926 days, except for the final interface which is shown at breakthrough for the* $10 \leq r \leq 200$ *simulation discussed in the text. Breakthrough occurs at 2,550 days. The measured breakthrough cone half-angle is approximately* $30°$. *Gravity dominated oil counterflow occurs for* $r \geq 135$ *ft. The r- and z-axes have been plotted to scale.*

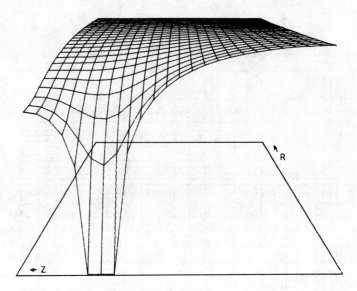

FIG. 14. *The breakthrough pressure field $P(r, z, t = breakthrough)$ for the run of Fig. 13. Although the calculation was done on a larger grid, for visibility of detail in reproduction the pressure has been plotted here on a 24×24 grid. Some small loss of detail around the outlet region results.*

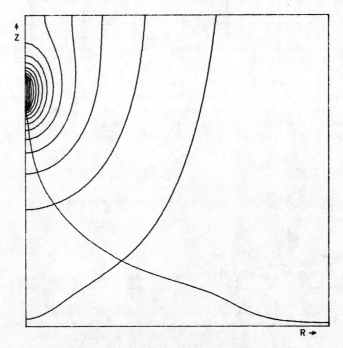

FIG. 15. *Fig. 14 drawn showing contours of constant pressure. The contours are equally spaced in pressure, the spacing being 43.75 psia.*

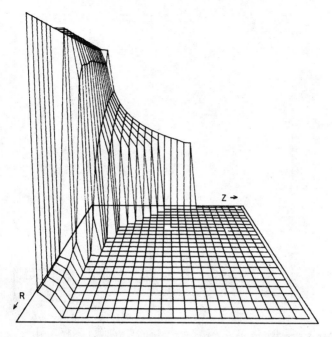

FIG. 16. *The breakthrough saturation profile for the run of Fig. 13. The oil rarefaction in the counterflow region $r \geq 135$ ft is clearly delineated. Although the calculation was done on a larger grid, for visibility of detail in reproduction the saturation has been plotted here on a 24×24 grid. Some small loss of detail results.*

plotted on a coarser 24×24 grid for clarity in reproduction. Some coarsening of detail results.) The shock height in the water bank is approximately $s = .41$, and in the counterflowing oil bank $s = .09$ (the exact height depending on the local gravity and velocity values).

Since our method of solving (2.1a) does not explicitly guarantee conservation of material, a check on the conservation law is an added diagnostic for the calculations. Fig. 17 shows plots of the accumulated horizontal material flow through the reservoir as a function of radius, namely

$$(6.1) \qquad \phi_i(r, t) = 2\pi r \int_0^t \int_{z_{min}}^{z_{max}} v_r(r, z, t) s_i(r, z, t) \, dz \, dt,$$

where i stands for the water or oil component. A maximum variation of about 5.6% is found in the total material flow ϕ_{tot} through the reservoir. A second check on the run is the enforcement of the no flow conditions at the outlet boundary. The outlet region is specified to lie in $35 \leq z \leq 40$. Diagnostics indicate that 90.7% of the material outflow was through this opening. However 97.5% of the material outflow occured through the opening $34.125 \leq z \leq 40.875$ in the $r = 10$ boundary. Thus the quality of the velocity (pressure) solution is to make the

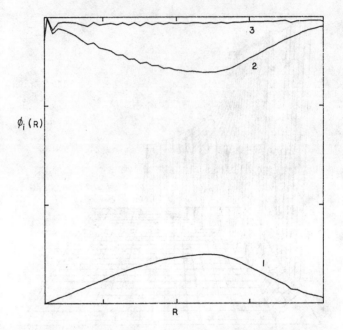

FIG. 17. $\phi_i(r, t = breakthrough)$ (equation (6.1)) for the run of Fig. 13. 1) i = water component. 2) i = oil component. 3) i = oil plus water. The maximum variation in curve 3 is 5.6%.

outlet slightly larger than specified (in this case, an increase from 10% of the total $r = 10$ boundary length to 13.5%).

Two more calculations were performed in order to obtain increased resolution of the cone region. Since the values $P(r = 100, z, t)$ in the $10 \leq r \leq 200$ run were found to remain constant in time at fixed z to within 0.4%, the values of $P(r = 100, z, t = 1852$ days$)$ from this run were used as boundary values for a run covering the cylinder $10 \leq r \leq 100$, $0 \leq z \leq 50$. All other reservoir and simulation parameters were the same, except that the grid used to solve equation (2.1a) was changed from 48×48 to 40×48. No significant loss of resolution accompanied this change which was implemented to lessen the total cpu time of the run (which ultimately depends on the ratio of the grid spacings in the r and z directions). Figure 18 plots the front at time intervals of 926 days except for the final front which is shown at breakthrough (2560 days). The measured cone half-angle for the breakthrough front is approximately 24°. The final run used the values $P(r = 50, z, t = 1852$ days$)$, which were observed to vary about 1% (at fixed z) over the course of the run, from the $10 \leq r \leq 200$ run as boundary values for a simulation covering the cylinder $10 \leq r \leq 50$, $0 \leq z \leq 50$. Fig. 19 plots the front for this run at time intervals of 926 days except for the final front which is shown at breakthrough (2480). The measured cone half-angle is approximately 15°.

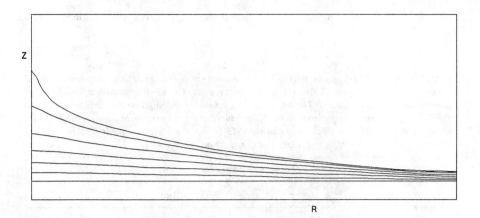

FIG. 18. *The developing cone shown at intervals of 926 days, except for the final interface which is shown at breakthrough for the $10 \leq r \leq 100$ simulation. Breakthrough occurs at 2,560 days. The measured breakthrough cone half-angle is 24°. The r- and z-axes are drawn to scale.*

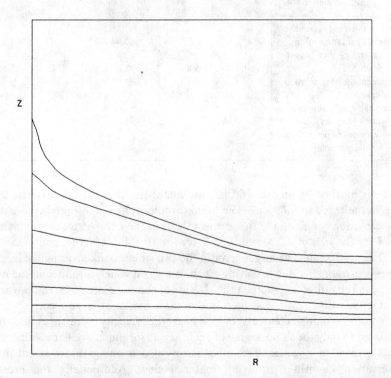

FIG. 19. *The developing cone shown at intervals of 926 days, except for the final interface which is shown at breakthrough for the $10 \leq r \leq 50$ simulation. Breakthrough occurs at 2,480 days. The measured breakthrough cone half-angle is 15°. The r- and z-axes are drawn to scale.*

Table 2 displays comparisons for quantities of interest among these three runs.

TABLE 2

Comparison of quantities of engineering interest among the three simulations. The lines marked with an asterisk should have identical values among the runs. These entries provide a further test of the accuracy of the calculation. The two quantities related to overall flow are consistent with the $\pm 0.8\%$ to $\pm 2.8\%$ error bars shown in the oil produced entry. (The $\pm$ error bars are obtained by comparing oil flux at the well opening and far field boundaries.) See text for a discussion of the measured cone half-angles.

	Run comparisons		
	Run		
	$10 \leq r \leq 200$	$10 \leq r \leq 100$	$10 \leq r \leq 50$
Total simulation volume (ft^3)	6,267,480	1,555,090	376,991
Initial oil present (as percentage of simulation volume)	90	90	90
Final oil present (as percentage of simulation volume)	91.5	82.94	75.99
*Breakthrough time (days)	2,550	2,560	2,480
*Oil produced (ft^3)	750,800 $\pm$ 21,500	764,400 $\pm$ 18,600	742,400 $\pm$ 5,660
As percentage of simulation volume	12	49	197
*Cone half-angle at breakthrough (degrees)	30	24	15
Cone volume (as percentage of simulation volume)	9.9	23	35

The lines marked by an asterisk indicate quantities that should have the same value in the three simulations. The breakthrough time and oil produced entries show consistency among the three runs to within their respective errors, namely $\pm 2.8\%$ for the $10 \leq r \leq 200$ run, $\pm 2.5\%$ for the $10 \leq r \leq 100$ run, and $\pm 0.8\%$ for the $10 \leq r \leq 50$ run. The larger error in the two simulations covering the larger r ranges is presumably due to solving (2.1b) on a grid which is refined in the mesh spacing in the r direction more at the r_{min} side of the calculational area than at the r_{max} side (see Fig. 12). Consequently the mesh blocks at large r correspond to much larger volume elements over which the velocity (pressure) has been calculated to only the same degree of accuracy as for much smaller volume mesh blocks at smaller r values. This error is revealed to the greatest extent in the conservation quantities involving material flow. Additionally the pressure decreases rapidly to the outlet in a much more narrow zone in the $10 \leq r \leq 200$ run. The decrease is much less rapid and less narrow (with respect to mesh spacings) in the $10 \leq r \leq 50$ run. Thus the maximum variation seen in the total

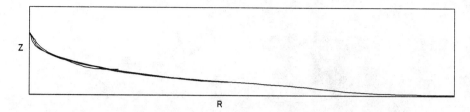

FIG. 20. *Superposition of the fronts for three coning simulations. The three runs are for* 1) $10 \leq r \leq 200$, 2) $10 \leq r \leq 100$, 3) $10 \leq r \leq 50$ *views of the same reservoir. Shown are the superposed fronts at breakthrough.*

material flow $\phi_{tot}(r)$ which is 5.6% in Fig. 17 and occurs near the $r = 10$ boundary, decreases to only 0.8% in the $10 \leq r \leq 50$ calculation.

Fig. 20 superposes the fronts for these three runs at their respective breakthough times. Comparison of these runs shows agreement for oil production and breakthrough times, but not for cone angle. We suspect the difference in cone angle is due to the extra refinement of the $10 \leq r \leq 50$ run which allows greater resolution of the narrowest part of the finger and thus a smaller cone angle. If this hypothesis is correct, we can conclude that the overall quantities such as oil produced have already converged in the $10 \leq r \leq 50$ run while certain details in the immediate neighborhood of the well bore have not. We note that to obtain a correct calculation in the immediate neighborhood of the well bore other effects should be included which modify the basic equations (2.1). The most prominent of these effects is the formation damage in the neighborhood of the well bore.

The reason for the discrepancy of the front in the region $r > 30$ between the $10 \leq r \leq 50$ run and the other two is unresolved as yet and requires further study.

We note further, that our use of $r = 10$ ft as a minimum r boundary does not reflect any restriction of our method. Smaller minimum r values (calculating the flow closer to the actual well bore) are possible. In particular, a run employing the same input values as the $10 \leq\ \leq 50$ run above, but with the inner boundary radius set to $r = 2.5$ ft, produced results for the cone shape (clearly the material flow quantities, oil produced, breakthrough time, etc. were different) that are in very good agreement with the $10 \leq r \leq 50$ run reported here. The measured cone half-angle in this $2.5 \leq r \leq 50$ run was approximately 18°.

As a final sidelight, Fig. 21 gives an example of an interesting way to graphically view a reservoir calculation. Plotted are nine fractional flow functions (2.2) calculated at different points in the breaktough field for the $10 \leq r \leq 200$ run. Their position in Fig. 21 *roughly* indicates the relative positions in the calculational area from which they were chosen. A generally increasing fractional flow curve indicates an upward field of flow, generally decreasing indicates a downward field of material flow. The absence of a minumum (or maximum) indicates the region is dominated by the pressure

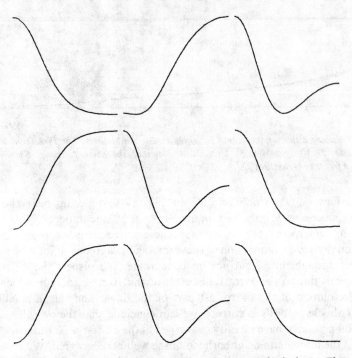

FIG. 21. *Examples of fractional flow curves encountered in an actual calculation. These examples were taken from the $10 \leq r \leq 200$ run at breakthrough. Their positions in the plot* roughly *indicate the relative positions in the reservoir from which they were computed.*

gradient. The presence of a minimum indicates the gravitational gradient is of comparable strength to the pressue gradient. The solution of a Riemann problem based on each of these consists in drawing the appropriate convex or concave envelope of the fractional flow curve.

7. Conclusions. The dependence on mesh orientation for conventional reservoir methods has been reported in the petroleum engineering literature [8], [9], [10]. Solutions adopted with some success have been choice of a favored coordinate system [9] or changes in a method to deal with the specific problem [10]. Excess numerical diffusion resulting from coarse grids and the desire to eliminate (numerical, but unphysical) saturation oscillations is another feature of conventional methods. All these problems are overcome by front tracking. We have demonstrated the high resolution ability of front tracking for the oil-water coning interface in vertical reservoir simulations. It is proposed to extend the method to the oil-gas coning problem and problems involving gas bubble percolation.

Acknowledgments. The work of James Glimm was supported in part by the National Science Foundation under grant MCS-8207965, and in part by the U.S. Department of Energy under grant DE-AC02-76ER03077. The work of Brent Lindquist and Oliver McBryan was also supported in part by the Department of Energy grant. McBryan is a Sloan Foundation Fellow.

REFERENCES

[1] J. GLIMM, D. MARCHESIN, O. MCBRYAN, AND E. ISAACSON, *A shock tracking method for hyperbolic systems,* in ARO Report 80-3, 1980.

[2] J. GLIMM, E. ISAACSON, D. MARCHESIN, AND O. MCBRYAN, *Front tracking for hyperbolic systems,* Adv. Appl. Math., 2 (1981), pp. 91–119.

[3] J. GLIMM, D. MARCHESIN, AND O. MCBRYAN, *A numerical method for two phase flow with an unstable interface,* J. Comp. Phys., 39 (1981), pp. 179–200.

[4] R. RICHTMYER AND K. MORTON, *Difference Methods for Initial Value Problems,* Wiley-Interscience, New York, 1967.

[5] A. O. GARDER, D. W. PEACEMAN, AND A. L. POZZI, *Numerical calculation of multidimensional miscible displacements by the method of characteristics,* Soc. Pet. Eng. J., 4 (1964), pp. 26–36.

[6] O. MCBRYAN, *Elliptic and hyperbolic interface refinement,* in Boundary Layers and Interior Layers—Computational and Asymptotic Methods, J. Miller, ed., Boole Press, Dublin, 1980.

[7] W. PROSKUROWSKI, *A note on solving the Buckley–Leverett equation in the presence of gravity.* J. Comp. Phys., 41 (1981), pp. 136–141.

[8] M. R. TODD, P. M. O'DELL, AND G. J. HIRASAKI, *Methods for increased accuracy in numerical simulation,* Soc. Pet. Eng. J., 12 (1972), pp. 515–530.

[9] G. E. ROBERTSON AND P. T. WOO, *Grid-orientation effects and the use of orthogonal curvilinear coordinates in reservoir simulation,* Soc. Pet. Eng. J., 18 (1978), pp. 13–19.

[10] J. L. YANOSIK AND T. A. MCCRACKEN. *A nine-point, finite difference reservoir simulator for realistic prediction of unfavorable mobility ratio displacement,* Soc. Pet. Eng. J., 18 (1978), pp. 253–262.

[11] J. GLIMM, E. ISAACSON, B. LINDQUIST, O. MCBRYAN, AND S. YANIV, *Statistical fluid dynamics: The influence of geometry on surface instabilities,* in this volume, Chapter IV.

[12] O. MCBRYAN, *Computing discontinuous flows,* in Proc. Meeting on Fronts, Patterns and Interfaces, Los Alamos, NM, 1983.

[13] J. GLIMM, O. MCBRYAN, R. MENIKOFF AND D. SHARP, *A shock-tracking method for studying Rayleigh–Taylor instability,* Los Alamos Technical Report, in preparation.

[14] J. GLIMM, O. MCBRYAN AND B. PLOHR, *Applications of gas dynamics to two-dimensional gas dynamics calculations,* in Lecture Notes in Engineering, Springer-Verlag, Berlin, 1983.

[15] O. MCBRYAN, *Multi-grid methods for discontinuous equations,* Proc. International Multi-grid Conference, Copper Mountain, CO, 1983 to appear.

[16] ———, *Shock tracking methods in 2-D flows,* Proc. 9th National Applied Mechanics Congress, Cornell University, Ithaca, NY, 1982.

CHAPTER IV

Statistical Fluid Dynamics: The Influence of Geometry on Surface Instabilities

JAMES GLIMM, ELI ISAACSON, BRENT LINDQUIST,
OLIVER McBRYAN, AND SARA YANIV

1. Introduction and results. Fingering is an instability of a fluid interface which occurs in a number of contexts. It was investigated for oil reservoirs initially by Taylor and Saffman [17] and later in other laboratory studies [2], [10], [15]. Related instabilities occur in astrophysics (surface instabilities of pulsating stars) [13], ionospheric irregularities [14], and the Rayleigh–Taylor instability associated with falling water drops [1], [12].

In a typical reservoir history, oil is first produced with reservoir pressures acting as the direct driving agent (primary production). Later, as decreased pressure gradients become unable to displace remaining oil, a recovery agent, commonly water, is injected into some of the wells to help displace the oil (secondary production). In the second case one finds oil-water discontinuities (shock waves) in the solutions to the simplest equations used to model this phenomenon. Often in the first case, water will be present initially in the reservoir, segregated from the oil by gravitational forces. Thus, in either case, there may be oil-water discontinuity interfaces.

A linearized, small amplitude analysis of disturbances to a planar oil-water interface reveals two regimes. The small amplitude fingers either grow (the unstable case) or decay (the stable case) with an exponential rate. For oil reservoir equations, the dimensionless number which governs the division between these cases is called the mobility ratio M defined to be the ratio of transmissibility of the fluid just behind the discontinuity, to the transmissibility of the fluid just ahead of the discontinuity. The reservoir equations and the meaning of transmissibility are given in the companion paper [9], Chapter III of this volume. The value $M = 1$ is the threshold for instability growth; small perturbations grow for $M > 1$ and disappear for $M < 1$ [4], [17]. Note that the mobility ratio defined here is called the "frontal mobility ratio" in the petroleum reservoir literature to differentiate it from a similar ratio involving fluid transmissibilities at points away from the discontinuity.

In the large amplitude regime the situation is more complicated, requiring the use of numerical computation to unravel the dependencies of the instabilities on M and other secondary effects. In earlier work we studied the development of a single finger [5], [7]. By a statistical method[1] we are able to study the large amplitude behavior of both single and multiple fingers [3], [5]. In an earlier study of the fully nonlinear theory, done in what we refer to as a linear geometry (a geometry not affected by sources or sinks), we found stable and unstable regimes lying below and above $M = 1$ respectively. Recent work by Chorin [18] indicates that there is a region $M_0 < M < 1$ in which instabilities in linear geometries will in fact grow if they are of sufficiently large initial amplitude. It is the increased transmissibility within the rarefaction wave behind the shock which drives the instability in this (linearly stable) regime $M < 1$. This indicates that a single dimensionless number such as M is insufficient to characterize the stable and unstable regimes.

Further studies have indicated the presence of other secondary phenonomena that come into play to modify or affect the onset of instability. Three such phenomena involved in the development of large amplitude fingers are Helmholtz instability along the finger sides, bifurcation of the finger tip and droplet formation (pinch-off). Observations both in the laboratory and in large amplitude calculations show that as a finger grows, its tip broadens and its base narrows. This is the early stage of a Helmholtz instability. As the instability grows the edge of the tip may roll over forming a mushroom cap—a later stage in Helmholtz instability. Or, the edges of the tip may move faster than the tip center leading to bifurcation of the finger. Finally the base of the finger may pinch off, and the finger become an isolated droplet. Although we observed these secondary phenomena to some extent in our calculations [5], [6], we have not studied them systematically. These secondary phenomena are sensitive to details of fluid modeling (surface tension, viscosity, inertial forces, . . .) and in our case, when all such stabilizing effects have been omitted from the equations, the secondary phenomena tend to destabilize the calculation.

Secondary phenomena which set the length scales in a reservoir include diffusion, capillary pressure and rock heterogeneity. These affect both the large and small scale structure of the displacement process. Some of these length scales serve to characterize the internal structure of the discontinuity. In mathematical terms, the displacement of the oil by the recovery agent proceeds with a sharp front (shock, or bank) followed by a smooth rarefaction wave. The shock is not perfectly sharp, but has internal structure of a diffusive character, caused by capillary pressure. For fixed rock properties, the length of the transition region internal to the shock is inversely proportional to the fluid velocity. As an approximation, we have omitted capillarity from the numerical results presented here; our calculated fronts are therefore perfectly sharp. The validity of the zero capillarity approximation depends of course on the problem being studied. In the present paper we model two-dimensional laboratory experiments where x-ray shadow graphs reveal that the internal shock structure is relevant only to very

---

[1]The basic idea being that the variance of a fingered interface around its mean position is used as a measure of finger length.

fine grid calculations [10]. Within the linearized, small amplitude theory, corrections for surface tension and small size effects (relevant to laboratory core flood experiments) have been studied [19]. See also [11] for stratification effects upon diffusion in a three-dimensional reservoir.

These length scales also determine the density of fingering formed in the unstable regime. In fact the length for formation of fingers should be the smallest length scale on which there is significant heterogeneity and which is larger than the diffusion or capillary length scale. (With diffusion and capillarity pressure terms absent from our calculation, the small scale limits on finger formation are set by numerical scales such as grid sizes.) Reservoir heterogeneity, by its very nature, is difficult to characterize in a calculation. We have studied the effects of normally distributed heterogeneity functions for areal reservoir calculations. The heterogeneity can be introduced either in the reservoir parameters [7] or in the initial discontinuity interface [6], [8]. In both cases we have shown that a solution containing multiple fingers is the result.

Heterogeneity can also be used in vertical studies to define geological stratification, thus allowing us to calculate channel flow near a producing well. We note that the general question of existence of fingering on different length scales in oil reservoirs appears unresolved within the petroleum engineering community [20, see especially p. 34, column 2, ¶2]. However the existence of channel flow, or fingering seen in vertical cross-section, caused by heterogeneous rock layers is not controversial. In Chapter III, we model vertical fingers resulting during primary production (the coning problem). The existence of these fingers is also not disputed.

Another secondary phenomenon revealed by both laboratory studies and calculations is competition between fingers. This can be viewed as fingering on a longer length scale; if equal fingers are modulated with a long wave length disturbance, some become slightly larger and others slightly shorter. The same instability which causes fingers also favors the long ones over the short ones and, as the differences in length increases, eventually only the long fingers grow significantly.

Calculations done in nonlinear geometries (i.e. near sources and sinks) indicate that geometry can play a role as important as mobility ratio in determining the stability of fluid interfaces. We present our results concerning the effect of geometry on fingering instabilities in this paper. In particular, we concentrate on the effects of both diverging (near an injection well) and converging (near a production well) geometries.

We use the incompressible two-phase flow equations as described in Chapter III for miscible and immiscible displacement. The frontal mobility ratio is defined as a ratio of transmissibilities

$$M = \frac{\lambda(s^-)}{\lambda(s^+)},$$

where s^+, s^- are the saturation values on opposite sides of the interface. In the miscible case we only consider the values $s^+ = 0$, $s^- = 1$. In this case the transmissibilities reduce to values of inverse velocity,

$$\lambda(s^+) = \mu_{oil}^{-1}, \qquad \lambda(s^-) = \mu_{solvent}^{-1},$$

and $M = \mu_{oil}/\mu_{solvent}$ becomes a viscosity ratio.

The case of immiscible displacement is not highly unstable. This includes the most common reservoir engineering methods, such as water flooding and gas injection. For a common Buckley–Leverett flux function [16], the frontal mobility ratio attains a maximal value of 2 as the viscosity ratio tends to infinity. In gas driven recovery, very large viscosity ratios are common. Mathematically, one can visualize the (near) stabilization of these flows as follows. The height of the shock (gas bank) is fairly small, so that the front consists of a jump between moderately similar fluids. The region ahead of the shock consists of oil with dissolved gas, the region behind the shock contains in addition a concentration of small bubbles of free gas. The frontal mobility ratio, by its definition, reflects this fact and as the viscosity ratio goes to infinity, the shock height goes to zero in such a way that the frontal mobility ratio remains bounded. We have mentioned that there are other (and more common) notions of mobility ratio, which become very large in the case of gas driven recovery, but which appear to have a less direct connection to questions of interface stability.

Miscible displacement has the potential for much greater instability. As we see from the above equation, the frontal mobility ratio can be given an arbitrarily large value. An example of miscible displacement in engineering practice is CO_2 flooding. It is recognized that fingering is an important aspect of the engineering design of a CO_2 flood. Normally, CO_2 injection is alternated with water to reduce fingering. A review of the engineering literature concerning unstable miscible displacement and its relevance to CO_2 flooding is given in [2].

1.1. Results for geometric studies. Although the front tracking method we employ is extendable to three-dimensional calculations, such calculations are beyond the current capabilities of our code. In any case it is desirable to use two-dimensional calculations where possible, to allow finer grid resolution and a more thorough exploration of the problem parameter space. We thus employ two-dimensional calculations, viewing the reservoir from above (areal view, x, y coordinates) or in vertical cross-section (cylindrical coordinates r, z). For the areal calculations, a typical reservoir flow geometry is given by the "quarter five-spot" problem, which is a square having Neumann (no flow through boundary) boundary conditions with a point source and sink respectively at two diagonally opposite corners. In this geometry, an initial interface is given as a quarter of a circle around the point source (injection well). As fluid is injected, the quarter circle grows in diameter. Gradually the influence of the point sink (production well) in the opposite corner is felt and, in simulations with no heterogeneity, a single finger is formed, pointing into the production well. We believe, for the case of no heterogeneity, that this single finger solution is stable in the mathematical sense against infinitesimal perturbations. Numerical calculations reported in [6] and Chaper III display this behavior. The presence of this single finger and its stability under both mesh refinement and change of grid orientation were reported in [6], even for the region $M \gg 1$. However this single finger solution is not observed experimentally [10] for $M \gg 1$, evidently because experimental sandbeds (as well as oil reservoirs) contain noninfinitesimal heterogeneities. We have shown that heterogeneities in the reservoir parameters [7] or in the initial reservoir interface [6], [8] lead to multiple fingers in the solution of this problem.

We idealize the five-spot problem somewhat by studying the flow field in the neighborhood of a single well (either producing or injecting). The injection well geometry was studied in areal view, the production well geometry from both areal and vertical views. We present our results below. Section 2 is then devoted to a detailed discussion of the injection well studies (diverging geometry) and §3 treats the producing well (converging geometry).

In the case of $M \leq 1$, the diverging geometry is absolutely stable. For $M > 1$, there is a stable ratio

$$(1.1) \qquad \frac{r_{\text{tip}}}{r_{\text{base}}} = R_d(M)$$

of finger tip to finger base positions (measured as distance to source) which is a function of the mobility ratio M achieved over the range of initial perturbation sizes used. In response to deterministic data (fingers in the initial interface) or statistical data (heterogeneous reservoir parameters) of varying intensity and frequency almost the same ratio $R_d(M)$ is observed. There is an approximate agreement between the numerical and laboratory experiments in this case.

For the converging geometry, the case $M \geq 1$ is absolutely unstable, while the case $M < 1$ however, is stable only for sufficiently small perturbations, having size smaller than a critical size

$$\frac{r_{\text{base}}}{r_{\text{tip}}} = R_c(M).$$

Above this critical size, perturbations grow.

A simple theory predicts a critical size for these unstable perturbations. This phenomena is not found in the experimental data of [10], but we have observed it in earlier numerical calculations (unpublished) as a numerically initiated instability.

Returning to a complete five-spot problem, we then expect the beginning of the calculation to follow the diverging geometry case and to be stabilized on geometrical grounds while the portion of the calculation near breakthrough of the front at the producing well will be destabilized, again on geometrical grounds.

The value of the frontal mobility ratio M which corresponds to engineering practice depends on the reservoir and the method of recovery. The most common injection fluid is water (secondary recovery) and typical values of M for a waterflood might be in the range .75 to 2.0. In this regime, we can conclude that finger length near the injection well, being limited by geometrical effects does not play a major role. In the case of CO_2 injection, a value M in the range 5 to 20 might be typical. However in order to reduce fingering, it is common to alternate slugs of CO_2 with slugs of water, so that a smaller effective M is achieved.

Near the producing well, fingering is more important. In the vertical view around production wells, fingering is further enhanced by the fact that heterogeneities occur in geological layers with fairly strong variation between distinct layers. Fingering associated with such layers is called channeling; a few high permeability layers will receive most of the flow, while the low permeability layers will be bypassed. The correct two-dimensional geometry for the study of channeling in the converging case near the producing well is therefore cylindrical, using r, z coordinates. The distribution of permeability between layers can be taken as log-normal (meaning that the exponents are normally distributed). Furthermore, in contrast to horizontal heterogeneity, layers can be included as deterministic data in the flow equations and do not have to be treated statistically. In fact the rock properties in the vicinity of a well can be determined quite accurately from sound speed and scattering measurements within the well bore.

Our channeling calculations are to some extent a simplification of real reservoirs. First, the layers we introduce are perpendicular to the well and parallel to each other (i.e. of uniform thickness). Secondly, geologically distinct layers with the thickness of centimeters do exist in reservoirs. The minimum length scale for fingering or channeling is set by a diffusion type term (capillary pressure) in the reservoir flow equations. This length scale is perhaps one or two orders of magnitude larger than these geological layers. Thus, there certainly exists significant geological variation on length scales too small to support channeling. As is stated in the literature [11] this variation increases the effective diffusion associated with the front. Since we do not include diffusion terms in our equations, the smallest length scales associated with our layers will set the length scales for fingering. However, the essence of the channeling problem is captured.

As a final comment, we note that production does not cease with breakthrough of the water interface at the production well. After breakthrough, a mixture of oil and water is produced and production continues until it is no longer economically practical to separate the oil from the water, a point which might be 95% water, 5% oil. Thus the slower channels will still produce oil after breakthrough if they are not too slow. All our fingering calculations were however terminated at breakthrough; fingering past breakthrough was not studied.

2. Diverging geometry.

2.1. Formulation of the problem.
In order to study a single well, we consider a quarter of an annulus bounded at radii r_1 and r_2 (see the boundary of Fig. 1a). The problem has Dirichlet boundary conditions on the two quarter circles[2] and Neumann conditions on the radial sides.

The equations of the two phase flow have the form

$$(2.1a) \qquad \frac{\partial s}{\partial t} + \mathbf{v} \cdot \nabla f(s) = 0,$$

$$(2.1b) \qquad \mathbf{v} = -\lambda(s)\nabla P, \qquad \nabla \cdot \mathbf{v} = 0,$$

where s denotes the water saturation, P is the pressure, $\mathbf{v}$ is the total fluid (oil plus water) velocity, $\lambda(s)$ the total fluid transmissibility and $f(s)$ the water fractional flow function. $\lambda(s)$ and $f(s)$ are known functions of saturation; in this chapter we use the functions describing miscible displacement (see [9]).

A conformal mapping $w = ie^{-iz}$ is used to map the problem from the annulus in the w-complex plane (the physical domain) onto a rectangle in the z-plane (henceforth called the computational domain). This gives rise to a mesh refinement in the radial direction.

The equations in the z-plane are given by

$$(2.2a) \qquad \frac{\partial s}{\partial t} + J \cdot \mathbf{v} \cdot \nabla f(s) = 0,$$

$$(2.2b) \qquad \mathbf{v} = -\lambda(s) \cdot \nabla P, \qquad \nabla \cdot \mathbf{v} = 0,$$

where $J = e^{-2y}$. Dirichlet boundary conditions are given along the lines $y = 0$, $y = Y$ and Neumann conditions are given along $x = 0$ and $x = \pi/2$.

In order to understand how multiple fingers behave, we consider a solution for the miscible case having fingers of equal length (i.e equal to $r_{\text{tip}}/r_{\text{base}}$) in the computational domain and uniform saturation of pure water behind, pure oil ahead of, the interface at time zero (see Fig.1a,b). We define $\lambda_2 = \lambda(1)$ and $\lambda_1 = \lambda(0)$ respectively as the pure water and pure oil transmissibilities. We compute analytically the rate of growth of the fingers using reasonable approximations, and compare with the numerical solution. We assume that the velocity field is vertical and given as in Fig. 1b. In the pure oil region the velocity is a constant v_3 and another constant v_1 between the fingers. In the pure water region

---

[2]The Dirichlet boundary conditions at the inner radius are used to model the presence of the injecting well. This has several obvious benefits, such as removing source terms from the RHS of (2.1).

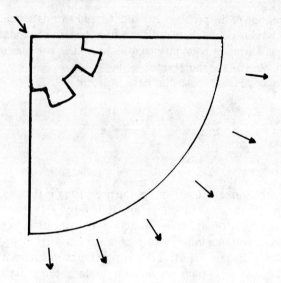

FIG. 1a. *Multiple fingers in a diverging geometry.*

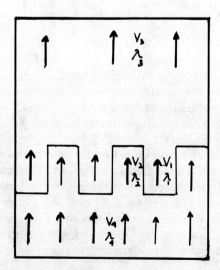

FIG. 1b. *Multiple fingers, velocity field and saturation front in computational geometry.*

the velocity inside each finger is a uniform v_2 and below the finger bases a constant v_4. Following the calculations of [5], we introduce the notation $(0, E) \equiv \mathbf{E} = -\nabla P$. Since $\nabla \times \mathbf{E} = 0$, a straightforward computation yields $E_1 = E_2$. As $v_1 = \lambda_1 E_1$, $v_2 = \lambda_2 E_2$, we have $v_2/v_1 = \lambda_2/\lambda_1$, hence the mobility ratio is the ratio of the velocities inside and outside the fingers.

We assume that the fingers and the region between the fingers have equal width. Since $\nabla \cdot \mathbf{v} = 0$, flux calculations yield $2v_3 = v_1 + v_2$ and $2v_4 = v_1 + v_2$. From the Dirichlet boundary conditions it follows that

$$P(x, y = 0, t) = P_\infty + \nu Y, \qquad P(x, y = Y, t) = P_\infty,$$

for some constant ν. We then obtain for an approximate solution of the pressure equation

$$\nu Y = E_3(Y - y_{\text{tip}}) + E_2(y_{\text{tip}} - y_{\text{base}}) + E_4 y_{\text{base}}.$$

Substituting for E_2, E_3, and E_4 in terms of λ_1, λ_2 and v_2 and solving for v_2 gives

$$(2.3) \qquad v_2 = \frac{2\nu\lambda_1\lambda_2}{\lambda_1 + \lambda_2 + (\lambda_1 - \lambda_2)(y_{\text{base}} - y_{\text{tip}})/Y}.$$

We observe as an experimental fact that fingers in the diverging cylindrical geometry assume a stable shape and grow under a similarity solution. This shape evidently depends on a variety of secondary causes (surface tension, diffusion layer, etc.) suppressed in our equations, and on the mobility ratio, M. In the computational geometry, the similarity transformation is a translation, $y \to y + \text{const}$. Once a stable shape is achieved, it is possible to relate this shape to the stable ratio $R_d(M)$ defined in (1.1). Suppose, for example, that the ideal shape of Fig. 1a were in fact the stable shape. Following the calculations of [5], we then obtain

$$(2.4) \qquad \frac{dy_{\text{tip}}}{dt} = e^{-2y_{\text{tip}}} v_2, \qquad \frac{dy_{\text{base}}}{dt} = e^{-2y_{\text{base}}} v_1,$$

and in the w-plane

$$(2.5) \qquad \frac{dr_{\text{tip}}/dt}{dr_{\text{base}}/dt} = \frac{r_{\text{base}}}{r_{\text{tip}}} M.$$

The fingers grow in the w-plane if

$$\frac{dr_{\text{tip}}}{dt} > \frac{dr_{\text{base}}}{dt}$$

and decay if

$$\frac{dr_{\text{tip}}}{dt} < \frac{dr_{\text{base}}}{dt}.$$

For $M \leq 1$, the fingers decay for any initial finger size. For $M > 1$, the diverging geometry stabilizes the fingering as follows: if the fingers at $t = 0$ are small, the factor $r_{\text{base}}/r_{\text{tip}}$ in (2.5) will be close to 1 and the fingers will grow until a stable length is reached, namely when

$$(2.6) \qquad R_d(M) = \frac{r_{\text{tip}}}{r_{\text{base}}} = M.$$

It can be observed that stable shapes will differ from block rectangular fingers and that (2.6) is not the correct value for $R_d(M)$ in general. In fact there is typically a broadening of the finger tip and a narrowing of the finger base (i.e. Helmholtz roll up along the finger sides). A study of these phenomena is outside the scope of the present paper, but certainly depends on surface tension and diffusion. It follows that the stable ratio R_d depends on surface tension and diffusion as well as on the mobility ratio.

As a slightly more complicated theory, we now define θ_{tip} to be the fractional (with respect to a length scale for the fingering pattern) finger width, measured at its widest part near the tip and θ_{base} to be the fractional finger width, measured at its narrowest part near the base. From the conservation laws, we obtain (for stable shapes)

$$\theta_{\text{tip}} \, v_{2,\text{tip}} = \theta_{\text{base}} \, v_{2,\text{base}} = M \, \theta_{\text{base}} \, v_{1,\text{base}},$$

where $v_{2,\text{tip}}$ and $v_{2,\text{base}}$ refer to velocities inside the finger at its widest and narrowest point respectively, and similarly for $v_{1,\text{tip}}$, $v_{1,\text{base}}$ between fingers. Substitution of these relations in (2.4) and (2.5) yields

$$(2.7) \qquad R_d(M) = \frac{r_{\text{tip}}}{r_{\text{base}}} = M \frac{\theta_{\text{base}}}{\theta_{\text{tip}}}.$$

This formula improves the fit to data, but we have not done sufficient tests to determine the true dependence of R_d on M and on other parameters such as finger shape.

Our main conclusions, tested over a variety of initial (single and multiple) finger sizes and widths, is that a unique, stable ratio $R_d(M)$ is observed. These tests were performed for the value $M = 5$ (see Table 1).

TABLE 1

Comparison among the stable ratios for M = 5 *and* M = 17.3 *for laboratory experiments and homogeneous and heterogeneous calculations. The ± error bars refer only to calculations actually performed. True error bars (from a larger set of calculations) could be larger.*

	$M = 4.58$	$M = 5$	$M = 17.3$
Experiment	1.42 ± 0.05		1.93 ± 0.13
Homogeneous, r_{tip}/r_{base}		1.45 ± 0.07	1.73
Homogeneous, $(r_{tip}/r_{base})_{eff}$		1.42 ± 0.07	1.66
Heterogeneous, r_{tip}/r_{base} physical length scales		1.43 ± 0.13	1.7
Heterogeneous, $(r_{tip}/r_{base})_{eff}$ physical length scales		1.42 ± 0.14	1.83
Heterogeneous, $(r_{tip}/r_{base})_{eff}$ computational length scales		1.65 ± 0.2	2.32

2.2. Statistical values for the finger length. In reality, the fingers have an irregular shape, different from the ideal shown in Fig. 1a. This is especially true in the case where heterogeneity in the reservoir parameters leads to multiple fingering in the solution (see §3). Thus, particularly in the heterogeneous case, the ratio R_d is not precisely defined. Rather, it must be interpreted as having meaning only in a statistical sense. Therefore, let mean(r) denote the mean value of a finger radius (in the physical (r, θ) geometry). This is a useful measure of finger height. Let variance(r) denote the variance of the finger radius from the mean. We shall relate the ratio variance(r)/mean(r) to r_{tip}/r_{base}.

Let us assume that the fingers in the computational region have the geometry shown in Fig. 1b. For convenience, we normalize the variable θ so that a single finger has width $\theta = 1$. With this convention, a finger in the (r, θ) domain is the step function

$$(2.8) \qquad f(r) = \begin{cases} r_{tip} & \text{if } 0 \leq \theta < \bar{\theta}, \\ r_{base} & \text{if } \bar{\theta} \leq \theta \leq 1, \end{cases}$$

where $\bar{\theta}$ is the fractional width of the finger (see Fig. 2).

Calculating the mean radius of the finger we obtain

$$\text{mean}(r) = r_{tip}\bar{\theta} + r_{base}(1 - \bar{\theta}),$$
$$\text{variance}(r) = (r_{tip} - r_{base})[\bar{\theta}(1 - \bar{\theta})]^{1/2}.$$

Hence,

$$(2.9) \qquad \frac{\text{variance}(r)}{\text{mean}(r)} = \frac{(r_{tip}/r_{base} - 1)[\bar{\theta}(1 - \bar{\theta})]^{1/2}}{(r_{tip}/r_{base} - 1)\bar{\theta} + 1},$$

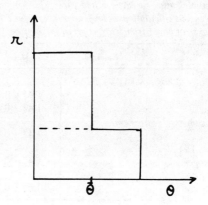

Fig. 2. *Typical finger in (r, θ)-plane*.

or

$$\frac{r_{\text{tip}}}{r_{\text{base}}} = 1 + \frac{K}{[\bar{\theta}(1 - \bar{\theta})]^{1/2} - K\bar{\theta}},$$

where $K = \text{variance}(r)/\text{mean}(r)$. From this analysis we define an effective ratio of $r_{\text{tip}}/r_{\text{base}}$ corresponding to R_d in the case of irregular fingers,

(2.10) $$\left(\frac{r_{\text{tip}}}{r_{\text{base}}}\right)_{\text{eff}} = 1 + \frac{K}{[\bar{\theta}(1 - \bar{\theta})]^{1/2} - K\bar{\theta}}.$$

Theoretical estimates of direct values for $r_{\text{tip}}/r_{\text{base}}$ and effective values obtained from the statistical variables (variance(r)/mean(r)) in (2.10) for an average width $\bar{\theta} = 0.5$ are given in Table 1.

We have calculated several flows in homogeneous reservoirs for $M = 5$. In each run a pattern of fingers of given equal width and amplitude were specified as initial data. The number of fingers, widths and amplitudes were however varied among the runs. An example of the fingering pattern at different times for one of the runs is shown in Fig. 3. Also runs for a homogeneous reservoir for $M = 17.3$ were done. Resulting values of measured $r_{\text{tip}}/r_{\text{base}}$ and $(r_{\text{tip}}/r_{\text{base}})_{\text{eff}}$ are shown in Table 1. Also included are measurements from the laboratory experiments of [10] and measurements from the heterogeneous reservoir calculations described in §3. As we see, there is agreement among these values; the fingers in each case reach a constant length though with slight differences among these constant lengths.

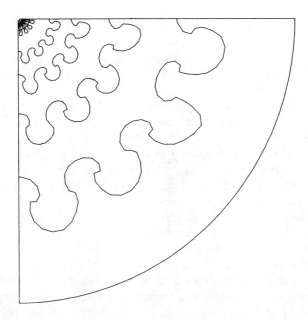

FIG. 3. *Miscible displacement, mobility ratio M = 5, homogeneous diverging geometry reservoir with given fingers in the initial front. The rate of growth of the fingers is constant and the dimensionless length r_{tip}/r_{base} depends only on M. It is independent of the amplitude and wave length of the initial perturbation.*

2.3. Fingering in heterogeneous media. In the previous section we discussed the stable ratio of finger length for deterministic data, i.e. fingers present in the initial interface. Typical oil reservoirs are heterogeneous, but since detailed information concerning the x, y dependence of the permeability is missing, we model heterogeneity as a random variable. The heterogeneity is introduced into the relative permeability by writing

$$\lambda = \lambda(s, x, y) = \lambda(s)\xi(x, y)$$

where $\xi(x, y)$ is a dimensionless random variable. To construct ξ, we introduce a grid with heterogeneity length scales l_x, l_y. For each lattice point

(2.11) $$x = il_x, \quad y = jl_y,$$

we construct a normally distributed random variable ξ_{ij} with mean 1 and specified variance. In order to avoid unphysical negative permeabilities, ξ_{ij} is then replaced by $|\xi_{ij}|$ at the points (2.11). We allow the grid with uniform spacing l_x, l_y to be constructed in either the physical w, or computational z coordinate system. In the former case the grid is referred to as having physical length scales, the latter as having computational length scales.

The reason for considering the grid of somewhat artificial computational length scales is the following: One can assume that real reservoirs are heterogeneous on all length scales. We make a simple attempt at modeling this possibility by allowing heterogeneity also on length scales measured in the computational coordinates which under the conformal remap $w = e^{-iz}$ covers a range of physical length scales. Also, presumably experimental sand beds from which data have been measured [10] are heterogeneous on length scales too small to be reasonably computed by available grid spacing. Our inclusion of a grid of physical length scales set somewhat larger than the mesh spacing allows some modelling of these very small length scales throughout the whole physical domain.

Now consider cylindrical coordinates r, z, and note that heterogeneity occurs in layers and so it is dominantly a function of z alone. To model the phenomena we further introduce $\xi = \xi(r, z) = \xi(z)$ a smooth interpolating function which agrees with ξ_k at $\xi = kl_z$ and is independent of r. In this case it is appropriate to take ξ_k log-normally distributed (so that log ξ_k is normal with mean 0 and given variance). No absolute value is required since $\xi_k > 0$ in this case.

Here we discuss the absolute normal heterogeneity computations for a diverging geometry. An example of fingers generated in a diverging geometry run with $M = 5$ and an absolute normal heterogeneity distribution of prescribed variance equal to 0.5 is presented in Fig. 4. The prescribed initial interface was

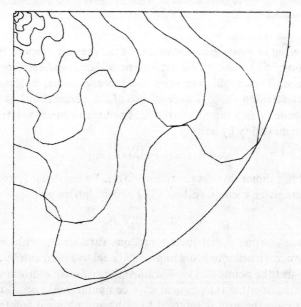

FIG. 4. *Miscible displacement, mobility ratio $M = 5$, heterogeneous diverging reservoir. The fingers are generated by heterogeneity with variance = 0.5 and physical length scales. The rate of growth of the fingers is constant as in Fig. 3.*

smooth. The results for r_{tip}/r_{base} are summarized in Table 1. Also shown are results from the laboratory experiments of [10]. The values of r_{tip}/r_{base} for the different homogeneous runs of the previous section, together with the heterogeneous runs for different variances of the heterogeneity random variable from this section are plotted in Fig. 5. These and the results of Table 1 show reasonable agreement among the experiment, homogeneous and heterogeneous (using physical length scales) runs.

The fingers with heterogeneity on computational length scales show similar but slightly more pronounced finger length. Heterogeneity on computational length scales, with variance in the range chosen here (e.g. variance(r) = 0.5) is an attempt to model residual or minor heterogeneity in a reservoir free from gross heterogeneity. Thus such a reservoir might, from an engineering point of view, be considered to be homogeneous. In addition to such engineering-homogeneous reservoirs, there are gross heterogeneities in certain reservoirs, which must be included explicitly to the extent that they are known. Examples of gross heterogeneities are: shale barriers, permeability streaks, fractures and channels. Only the latter of these possibilites, channels, will be considered in this paper (§3).

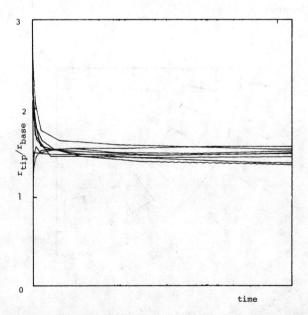

FIG. 5. r_{tip}/r_{base} *for mobility ratio* $M = 5$. *Curves are plotted for several homogeneous and heterogeneous reservoirs in a diverging geometry.*

3. Converging geometry. In the converging geometry for the areal problem near the producing well the conformal mapping $w = ie^{-iz}$ maps the physical w-domain to the rectangular z-region. In this case the Jacobian of the transformation is $J = e^{2y}$.

Following the analysis of §2, we obtain

$$(3.1) \qquad \frac{dy_{\text{tip}}/dt}{dy_{\text{base}}/dt} = \frac{e^{2(y_{\text{tip}} - y_{\text{base}})}}{R_c(M)}$$

for some function $R_c(M)$. Since $y = -\log r$, the fingers in the w-plane will grow (decay) if

$$\frac{dy_{\text{tip}}/dt}{dy_{\text{base}}/dt} > 1 \quad (<1),$$

or

$$\frac{dr_{\text{tip}}/dt}{dr_{\text{base}}/dt} = \frac{r_{\text{base}}/r_{\text{tip}}}{R_c(M)}.$$

In the stable case ($M < 1$), if we start with small fingers, the factor $r_{\text{base}}/r_{\text{tip}}$ will be close to 1 (in this geometry $r_{\text{base}} > r_{\text{tip}}$ as is shown in Fig. 6), and the fingers will decay. Fingers larger than the critical value $r_{\text{base}}/r_{\text{tip}} = R_c(M)$ will increase in time. In the unstable case ($M > 1$, $R_c(M) < 1$) the fingers will grow for every initial finger length.

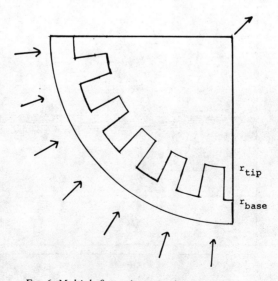

FIG. 6. *Multiple fingers in converging geometry.*

Repeating the previous analysis for the vertical converging geometry calculation reveals the same behavior for the fingers.

The flow equations are

$$\frac{\partial s}{\partial t} + \mathbf{v} \cdot \nabla_{(r,z)} f(s) = 0,$$

$$\mathbf{v} = - \lambda(s) \cdot \nabla_{(r,z)} P,$$

$$\nabla_{(r,z)} \cdot \mathbf{v} = \frac{1}{r}\frac{\partial r v_r}{\partial r} + \frac{\partial v_z}{\partial z} = 0.$$

The effect of the converging geometry on fingering has been examined by numerical calculations. We calculate heterogeneous reservoirs with heterogeneity taken along channels as described in §2.3. The heterogeneity function is illustrated in Fig. 7 where z is plotted vs. $\xi(z)$. This heterogeneity function creates channels in the lower part of the region along which the permeability function varies by almost a factor of 10. This variation is not at all extreme for real reservoirs. The section of the reservoir included in these calculations is 360^3 feet

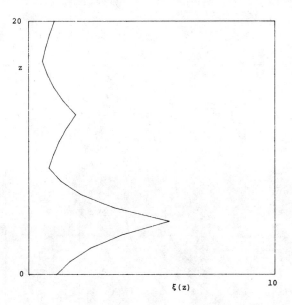

FIG. 7. *Channeling heterogeneity function. The vertical coordinate z is plotted vs. the heterogeneity function $\xi(z)$.*

---

[3] From $r_{min} = 40$ feet to $r_{max} = 400$ feet.

long and 20 feet deep. The grids we used had either 30 points in the r direction and 20 or 30 points in the z direction. See Chapter III for a discussion of logarithmic weighting used for the r direction grid lines in the solution of the elliptic equation. We consider both stable and unstable runs for miscible and immiscible displacement (waterflood). The stable case for miscible displacement has a mobility ratio $M = 0.5$ and the unstable mobility ratio is $M = 2$. For the immiscible displacement the calculations have been performed for a viscosity ratio of $\mu_{oil}/\mu_{water} = 2$ in the stable case, for which the mobility ratio is .85, and a viscosity ratio of $\mu_{oil}/\mu_{water} = 10$ in the unstable case, for which the mobility ratio is 1.4. In all cases the channel generated by heterogeneity grows toward the producing well until breakthrough is reached. The results for these problems have the same qualitative behavior; the differences in the displacement function or the mobility ratio do not affect the instabilized growth of the finger. The channel pattern at different time steps in the miscible displacement problem, is shown in Figures 8 and 9 respectively for the stable and unstable problems. In the immiscible displacement case the channel pattern is plotted in Fig. 10 for $\mu_{oil}/\mu_{water} = 2$ and in Fig. 11 for $\mu_{oil}/\mu_{water} = 10$. In Figs. 12–15 we show the velocity field for a typical time step in these four runs.

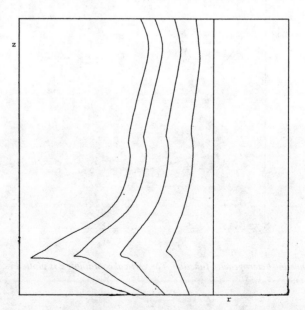

FIG. 8. *Front position at selected time steps in a converging geometry. The heterogeneity function is plotted in Fig. 7. The mobility ratio is $M = .5$ for miscible displacement.*

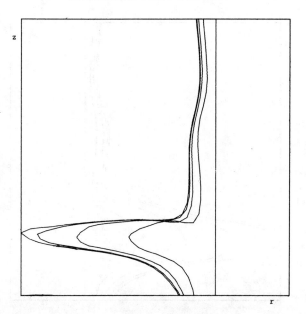

FIG. 9. *Front position at selected time steps in a converging geometry. The heterogeneity function is plotted in Fig. 7. The mobility ratio is $M = 2$ for miscible displacement.*

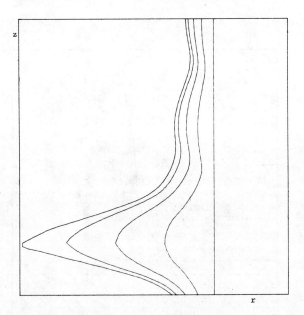

FIG. 10. *Front position at selected time steps in a converging geometry. The heterogeneity function used for this run is plotted in Fig. 7. The viscosity ratio $\mu_{oil}/\mu_{water} = 2$ for immiscible displacement.*

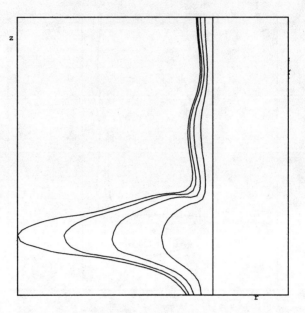

FIG. 11. *Front position at selected time steps in a converging geometry. The heterogeneity function used for this run is plotted in Fig. 7. The viscosity ratio $\mu_{oil}/\mu_{water} = 10$ for immiscible displacement.*

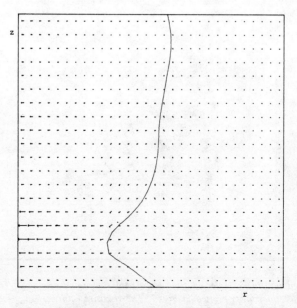

FIG. 12. *The velocity field for a typical time step for the case shown in Fig. 8.*

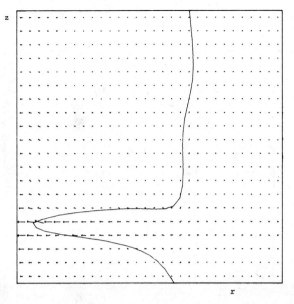

FIG. 13. *The velocity field for a typical time step for the case shown in Fig. 9.*

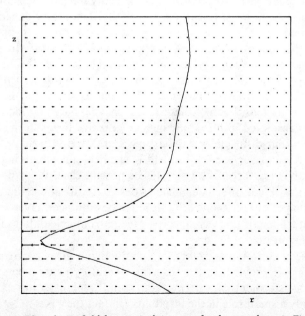

FIG. 14. *The velocity field for a typical time step for the case shown in Fig. 10.*

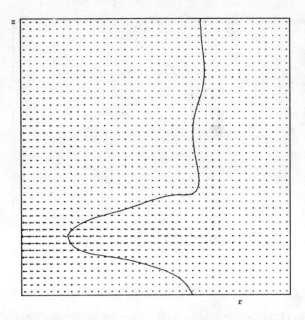

FIG. 15. *The velocity field for a typical time step for the case shown in Fig.* 11.

4. Conclusions. We have studied systematically some of the factors affecting the length of fingers in petroleum reservoirs and laboratory experiments. The factors studied were geometry (converging vs. diverging or production vs. injection well), heterogeneity and (frontal) mobility ratio. In the case of diverging geometry (an injection well) we conclude that the phenomena of fingering is not important. In this case we find excellent qualitative agreement and reasonable quantitative agreement between experimental data [10] and numerical calculations. Each of these approaches supports our main conclusion. Factors not considered in this study include diffusive terms in the equations of motion (molecular diffusion or capillary pressure) and subgrid or small length scale heterogeneity. Surface tension, which occurs in some fingering problems as a surface stabilizing force, is missing in the present case. We expect that these extra factors will not affect our conclusions, but will affect finger initiation, width and shape. In fact the width of the viscous diffusion layer in the data of [10] is about 5% and if this were incorporated into our calculations it would only reduce the importance of fingers.

In the case of converging geometry (a production well), we have attempted to model a reservoir problem rather than the laboratory experiments. For typical waterflood displacement parameters the (frontal) mobility ratio M is close to 1 and plays only a small role in the instability. In fact the cases $M < 1$ and $M > 1$

are not greatly different in area flooded. M is closely correlated to the shape of the fractional flow function $f(s)$ and thus to the height of the oil-water bank. Large values of M correlate with small saturation shock heights and in this way to lower production for the same area flooded. The instability is caused by the geometrical effect of convergence and is initiated by the fairly strong heterogeneity in distinct horizontal layers (channels) which one can typically expect in a real reservoir. The (omitted) diffusion terms in the equations of motion would still have the effect of reducing the importance of channeling and thus should be included in a further study of this problem. All our runs were stopped at breakthrough and a further study of this problem should include the production obtained after breakthrough. Our vertical calculations indicate that channeling is important and is governed by strong heterogeneity in horizonal layers, or channels, near the production well. This conclusion is qualified by factors omitted from the present analysis (diffusion and production after breakthrough) which are qualitatively expected to operate in the opposite direction.

Our main conclusion is that factors strongly influencing instability (geometry and heterogeneity) cannot be deduced from a study of the linear geometry (flow in a source/sink free region, or equivalently in a two-dimensional study, flow from a line source to a line sink). For example, neither the stability of the $M > 1$ diverging flows not the instability of the $M < 1$ converging flows can be predicted from the behavior of the linear geometries. Thus in a problem containing both geometry and mobility ratios as parameters, it is possible for the geometry to be the more important of the two.

Acknowledgments. The work of James Glimm and Sara Yaniv was supported in part by the National Science Foundation under grant MCS-8207965, and in part by the U.S. Department of Energy under grant DE-AC02-76ER03077. The work of Brent Lindquist and Oliver McBryan was also supported in part by the Department of Energy grant. McBryan is a Sloan Foundation Fellow.

REFERENCES

[1] G. R. BAKER, D. I. MEIRON AND S. A. ORSZAG, *Vortex simulations of the Rayleigh–Taylor instability,* Phys. Fluids, 23 (1980), pp. 1485–1490.
[2] E. L. CLARIDGE, *Prediction of recovery in unstable miscible flooding,* Soc. Pet. Eng. J., 12 (1972), pp. 143–155.
[3] J. GLIMM, D. MARCHESIN AND O. MCBRYAN, *Statistical fluid dynamics: unstable fingers,* Comm. Math. Phys., 74 (1980), pp. 1–13.
[4] ———, *Unstable fingers in two phase flow,* Comm. Pure Appl. Math., 34 (1981), pp. 53–75.
[5] ———, *A numerical method for two phase flow with an unstable interface,* J. Comp. Phys., 39 (1981), pp. 179–200.
[6] J. GLIMM, E. ISAACSON, D. MARCHESIN AND O. MCBRYAN, *Front tracking for hyperbolic systems,* Adv. Appl. Math., 2 (1981), pp. 91–119.

[7] J. GLIMM AND O. MCBRYAN, *Front tracking for hyperbolic conservation laws*, ARO Report 81-3, Proc. 1981 Army Numerical Analysis and Computer Conference.
[8] J. GLIMM, *Singularities in fluid dynamics*, in Mathematical Problems in Theoretical Physics, Proceedings, 1981, R. Schrod and D. A. Uhlenbrock, eds., Lecture Notes in Physics 153, Springer-Verlag, New York, 1982.
[9] J. GLIMM, B. LINDQUIST, O. MCBRYAN AND L PADMANABHAN, *A front tracking reservoir simulator, five-spot validation studies and the water coning problem*, this volume, Chapter III.
[10] B. HABERMANN, *The efficiency of miscible displacement as a function of mobility ratio*, Trans. AIME, 219 (1960), pp. 264–272.
[11] L. W. LAKE AND G. J. HIRASAKI, *Taylor's dispersion in stratified porous media*, Soc. Pet. Eng. J., 18 (1978), pp. 459–468.
[12] R. MENIKOFF AND C. ZEMACK, *Rayleigh–Taylor instability and the use of conformal maps for ideal fluids*, J. Comp. Phys., submitted.
[13] M. L. NORMAN, L. SMARR, K-H WINKLER AND M. D. SMITH, *Structure and dynamics of supersonic jets*, Preprint MPA 6, Max-Planck-Institut für Physik und Astrophysik, March 1982.
[14] S. L. OSSAKOW, M. J. KESKINEN AND S. T. ZALESAK, *Ionospheric irregularity physics modeling*, AIAA 20th Aerospace Sciences Meeting, Orlando, FL, 1982.
[15] L. PATERSON, *Radial fingering in a Hele–Shaw cell*, J. Fluid Mech., 113 (1981), pp. 513–529.
[16] A. SPIVAK, H. S. PRICE AND A. SETTARI, *Solution of the equations for multidimensional two-phase, immiscible flow by variational methods*, Soc. Pet. Eng. J., 17 (1977), pp. 27–41.
[17] S. I. TAYLOR AND P. G. SAFFMAN, *A note on the motion of bubbles in a Hele–Shaw cell and porous medium*, Quart. J. Mech. Appl. Math., 12 (1959), pp. 265–279.
[18] A. J. CHORIN, *The instability of fronts in a porous medium*, Univ. California preprint, Berkeley, CA, 1983.
[19] E. PETERS AND D. FLOCK, *The onset of instability during two phase immiscible displacement in porous media*, Soc. Pet. Eng. J., 21 (1981), pp. 249–258.
[20] F. CRAIG, *The reservoir engineering aspects of waterflooding*, American Institute of Mining, Metallurgical and Petroleum Engineers, Dallas, 1971.

CHAPTER V

Some Numerical Methods for Discontinuous Flows in Porous Media

PHILLIP COLELLA, PAUL CONCUS, AND JAMES SETHIAN

1. Introduction. The numerical modeling of fluid displacement through a porous medium has received increased attention in recent years. Interest has been stimulated by the development of enhanced recovery methods for obtaining petroleum from underground reservoirs and the advent of larger, higher-speed computers. A common feature found in most important recovery methods is the propagation of fronts that are steep or discontinuous. Examples of such fronts are those involving different fluids, such as in the waterflooding of a petroleum reservoir, or between regions of differing concentrations, as in some tertiary recovery processes. Even though steep fronts may not be present initially, they can develop naturally in time as a consequence of the inherent nonlinearity of fluid displacement in porous media.

Accurate following of steep fronts numerically can pose substantial difficulty for conventional discretization methods, which rely on underlying assumptions on smoothness of solutions. In an attempt to overcome these difficulties, a study was initiated several years ago in the Mathematics Group of the Lawrence Berkeley Laboratory to develop high-resolution numerical methods for solving the equations of flow through a porous medium. A discussion of this study is given here with emphasis on the details of newer directions being pursued. Included as well is some introductory background material given in an earlier, less-detailed review [14].

Our study centers on numerical methods that incorporate analytical information concerning the propagation of discontinuities in a flow. Such methods have been effective in treating hyperbolic conservation laws arising in gas dynamics and can be adapted in many cases to the equations of porous flow. The initial parts of the study focused on the random choice method, a method that can track solution discontinuities sharply and accurately in one space dimension. The method represents a solution by a piecewise constant approximation and uses Riemann problem solutions and a sampling procedure to advance in time. As a first step this method was adapted to solving the Buckley–Leverett equation for immiscible displacement in one space dimension. Extensions to more than one space dimension were carried out subsequently by means of operator splitting (§3).

Because of inaccuracies that may be introduced for some problems at discontinuity fronts propagating obliquely to the splitting directions, investigations of alternatives were initiated for multidimensional cases. A front tracking method for multidimensional problems was developed based on the SLIC scheme introduced in [26]. The method assigns to mesh cells a value representing the fraction of the cell lying behind the front, and the cell fractions are then appropriately advanced at each time step (§4). Two other methods, currently under investigation, are a higher order version of Godunov's method that utilizes piecewise linear rather than piecewise constant segments for constructing conservative fluxes (§5), and a method based on an improved splitting procedure, which will be reported in [9].

2. Equations for immiscible displacement. We consider the simultaneous, immiscible flow of two incompressible fluids in an isotropic, homogeneous porous medium. We do not include the effects of capillary pressure, thus propagating fronts will be sharp. For a region whose interior is free of sources and sinks (i.e., injection or production wells), one is led to the equations [28]

$$(2.1) \quad \phi \frac{\partial s}{\partial t} + \mathbf{q} \cdot \nabla f(s) - \gamma \frac{\partial}{\partial z} g(s) = 0,$$

$$(2.2) \quad \nabla \cdot \mathbf{q} = Q,$$

$$(2.3) \quad \mathbf{q} = -\lambda(s)[\nabla p - \gamma \tilde{g}(s)\mathbf{e}_k].$$

In the above equations $s(\mathbf{x}, t)$, $0 \leq s \leq 1$, is the saturation of the wetting fluid (fraction of available pore volume occupied by the fluid). The saturation of the nonwetting fluid is then $1 - s$. The independent variables $\mathbf{x}$ and t are space and time, respectively, and $\mathbf{q}(\mathbf{x}, t)$ is the total velocity (sum of the individual velocities of the two fluids). If gravity is present it is assumed to act in the negative z direction, with $\mathbf{e}_k$ the unit vector in the positive z direction. The quantity $p(\mathbf{x}, t)$ is the excess over gravitational head of the reduced pressure; here the reduced pressure is the average of the individual phase pressures less the gravitational head. The quantity Q represents the sources and sinks of fluid on the boundary of the domain, and ϕ is the porosity, which will be assumed constant. The quantity γ, the coefficient of the gravitational term, is the product of the acceleration due to gravity times the density difference between the wetting and nonwetting phases.

Equation (2.1) is the Buckley–Leverett equation, which for a given $\mathbf{q}$ is hyperbolic. Equation (2.2) is the incompressibility condition, and (2.3) is Darcy's law. For a given s, (2.2), (2.3) is elliptic.

The functions of saturation $f(s)$, $g(s)$, $\lambda(s)$, and $\tilde{g}(s)$ can be expressed in terms of the empirically determined phase mobilities (ratios of permeability to viscosity) λ_n and λ_w of the nonwetting and wetting fluids. For immiscible displacement these are

$$f(s) = \lambda_w/\lambda, \qquad g(s) = \lambda_n f,$$
$$\tilde{g}(s) = \lambda_n/\lambda, \qquad \lambda(s) = \lambda_n + \lambda_w.$$

The quantities f, g, and $\tilde{g}$ are nonnegative, and λ is positive.

A distinguishing feature of the immiscible displacement equations is that f and g are nonconvex. Typically f has one inflection, as depicted for a model case in Fig. 1, and g has two, as depicted in Fig. 2. Thus weak solutions may have combinations of propagating shock and expansion waves in contact.

Attempts to solve (2.1), (2.2), (2.3), subject to appropriate boundary conditions, by standard discretization methods such as finite difference or finite element methods can give rise to substantial difficulty. Inaccuracies may arise near a moving front, or an incorrect weak solution may be obtained. To circumvent these difficulties, the first phase of our study initiated an attempt to adapt the random choice method to solving problems of fluid displacement in porous media.

3. Random choice method. The random choice method, which was formulated originally for solving the equations of gas dynamics, is a numerical method incorporating the accurate propagation of solution discontinuities. It is based on a mathematical construction of Glimm [17] that was developed into a practical

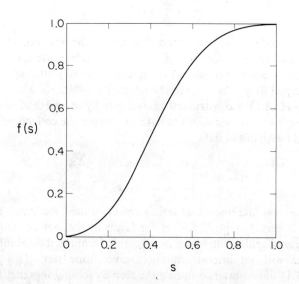

FIG. 1.

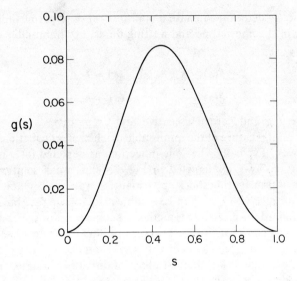

Fig. 2.

and efficient computation algorithm by Chorin [6], [7]. It was first adapted to porous flow problems in [15], and applied extensively in [18] and related papers.

For a single nonlinear conservation law

$$(3.1) \qquad \frac{\partial s}{\partial t} + \frac{\partial}{\partial z}\psi(s) = 0,$$

to which (2.1) reduces in one space dimension, the random choice method advances a solution in time as follows. The solution $s(z, t_n)$ at time t_n is represented by a piecewise constant function on a spatial grid of spacing Δz, with the function equal to $s_i^n = s(z_i, t_n)$ in the interval $z_i - \frac{1}{2}\Delta z < z \leq z_i + \frac{1}{2}\Delta z$. An exact solution of (3.1) is constructed analytically by the method of characteristics for this piecewise constant initial data by solving the collection of Riemann problems: (3.1) with initial data

$$(3.2) \qquad s(z, t_n) = \begin{cases} s_i^n, & z \leq z_i + \frac{1}{2}\Delta z, \\ s_{i+1}^n, & z > z_i + \frac{1}{2}\Delta z. \end{cases}$$

As long as the time increments Δt satisfy the Courant–Friedrichs–Lewy condition $(\Delta t/\Delta z) \cdot \max|\psi'(s)| < \frac{1}{2}$ (or <1 for forms of the method using half time steps on staggered grids), the waves propagating from the individual mesh-point discontinuities will not interact during a given time step. This permits the solution of (3.1) to be obtained during the step by joining together the separate Riemann problem solutions.

The above technique of obtaining the exact solution is common to other methods, such as Godunov's method. What distinguishes the random choice method is that the piecewise constant representation of the solution s is constructed at the new time by sampling the exact solution at a point within each spatial interval. In this way moving discontinuities remain perfectly sharp (since no intermediate values are introduced by the method), at the price of introducing a small amount of uncertainty into the position of the waves.

The sampling procedure for the random choice method should be equidistributed to yield an accurate representation of the solution [7], [10]. The deterministic van der Corput sequence proposed in [10] has been found to be particularly well suited for the method. The mth number θ_m in the basic sequence is given by

$$\theta_m = \sum_{k=0}^{M} i_k 2^{-(k+1)},$$

where the binary expansion for m is

$$m = \sum_{k=0}^{M} i_k 2^k.$$

Extensions for use with multidimensional problems are given in [10].

The random choice method is essentially first order and is observed to give good results for one-dimensional problems.

3.1 Riemann problems. The practicality of the random choice method depends on being able to solve the Riemann problems efficiently. For the immiscible displacement problem the function $\psi(s)$, which is a linear combination of $f(s)$ and $g(s)$, has either one or two inflections, depending on the relative magnitudes of q and γ. If the gravity term $\gamma g(s)$ in (2.1) is small compared with the transport term $qf(s)$, then there is only one inflection in $\psi(s)$ (as in $f(s)$ in

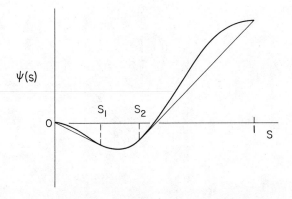

Fig. 3.

Fig. 1), for which case the Riemann problem solution is given in [15]. For the case of two inflections in $\psi(s)$ the solution is given in [1], [4], and for a special case in [29].

A typical example for which two inflections occur is depicted in Fig. 3, which is taken from [4]. The Riemann problem solution is obtained by applying the following general conditions, which must hold along any curve of discontinuity of $s(z, t)$: Let $s_- = \lim_{z \to z_-} s(z, t)$ and $s_+ = \lim_{z \to z_+} s(z, t)$ be the limiting values from the left and right, respectively, at a discontinuity. Then there must hold (see [24], [27]):

(i) *Rankine–Hugoniot jump condition.* The curve of discontinuity is a straight line with slope

$$\frac{dz}{dt} = \frac{\psi(s_+) - \psi(s_-)}{s_+ - s_-}.$$

(ii) *Generalized entropy condition.* For any s between s_+ and s_- there holds

$$\frac{\psi(s_+) - \psi(s)}{s_+ - s} \leq \frac{\psi(s_+) - \psi(s_-)}{s_+ - s_-}.$$

For the case $s_i^n = 0$ and $s_{i+1}^n = 1$, one obtains the solution of (3.1), (3.2) depicted in Fig. 4. Figure 3 depicts the corresponding concave hull of $\psi(s)$, whose points of tangency with $\psi(s)$ determine the shock propagation speeds. The two shocks shown in Fig. 4 propagate to the left and right, respectively, from the initial discontinuity. The characteristics from the left of the discontinuity intersect the leftward travelling shock, and those from the right intersect the

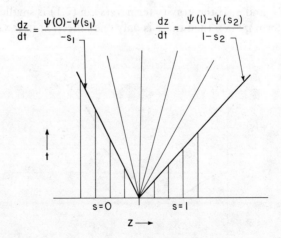

FIG. 4.

rightward travelling shock. Between the two shocks lies an expansion wave, whose fan of characteristics emanates from the initial discontinuity.

Further discussion of the Riemann problem for these equations can be found in [1] and [4].

3.2. Operator splitting for multidimensional problems. For a one-dimensional problem, the system (2.1), (2.2), (2.3) reduces to the single conservation law of the form (3.1), since, in this case, **q** is constant in the interior. For a multidimensional problem, a straightforward technique for solving (2.1), (2.2), (2.3) is to solve successively (2.2), (2.3) for p (and **q**) taking s to be fixed at its approximate solution for the current time, and then to advance (2.1) one time step considering **q** fixed, to obtain an approximate solution for s at the new time.

Advancing (2.1) is carried out using the random choice method in [1], [2], and [4] by means of operator splitting. Specifically, for two space dimensions, one solves successively the one-dimensional problems for s

$$\phi \frac{\partial s}{\partial t} + q_x \frac{\partial}{\partial x} f(s) = 0,$$

$$\phi \frac{\partial s}{\partial t} + q_z \frac{\partial}{\partial z} f(s) - \gamma \frac{\partial}{\partial z} g(s) = 0.$$

where $\mathbf{q} = (q_x, q_z)$. Numerical examples can be found in [1], [2], and [4].

Although this technique is efficient and gives acceptable results for many problems of interest, it can be inaccurate for some cases in which a shock front is advancing obliquely to the splitting directions (for example, see [16]). A modification of the split random choice method currently under development overcomes these difficulties and has been shown to give very good results for two space dimensions [9].

4. Front tracking in more than one dimension: SLIC. If one is interested in following the motion of a front in more than one dimension, several techniques are available. One method, mentioned above, is to use the technique of operator splitting to solve appropriate one-dimensional problems along each of the splitting directions in succession until the full multidimensional solution is achieved. In this technique, the location of the front, as such, is never used explicitly. Instead, the movement of the front is implicitly contained in the values of particular variables, in our case, the saturation.

Alternatively, one could choose to concentrate on the line of discontinuity and follow its motion in time. A standard technique is to spread a collection of marker particles along the front at the initial time in such a way that interpolation through these points provides a good approximation to the initial line of discontinuity. From this approximation to the front and the advection velocity provided by the hyperbolic equations being solved, the direction and speed of the front at each marker particle are determined. Each particle is moved, and interpolation provides the position of the front at the updated time. Of course, the success of the technique relies heavily on the accuracy of the interpolation. Since the direction of motion for each marker particle is obtained, in part, from the orientation of the front at that point, small errors in this orientation can lead to substantial error in the position of the front. In addition, as the marker particles move, they can spread apart as well as bunch together, and it is not a simple task to provide an accurate interpolation to the position of the front from such a collection of points. A discussion of some of the problems inherent in these methods when applied to flame propagation may be found in [31].

As a third alternative, a front tracking method can be based on a "volume of fluid" construction, such as the Simple Line Interface Calculation (SLIC) developed in [26]. In this technique, a grid is imposed on the domain and each cell is assigned a number corresponding to the fraction of that cell located behind the front. These cell fractions are updated during each time step, in accordance with the appropriate differential equations. The position of the front is approximated by a local construction in each cell, based on neighboring cell fractions. This technique does not rely on a global interpretation of the front, and thus can be highly effective in situations in which the front contains fingers and cusps. This method of tracking moving discontinuous fronts is an integral part of a flame propagation algorithm developed in [8], shown to be a natural construction from the point of view of the theory of flame propagation in [31], and used with highly successful results in [32] to model turbulent combustion. Such a method was used in [25] to follow shock discontinuities in Burgers' equation and the equations for two-phase porous flow. In the rest of this section, we discuss SLIC and the application of this type of front tracking method to petroleum reservoir simulations.

4.1. The method. We wish to follow the motion of a front in two space dimensions and assume for now that the velocity at the front is known at all times. We impose a square grid $\{i, j\}$ of uniform mesh size on the domain, and assign a number $0 \leq f_{i,j} \leq 1$ to each cell, corresponding to the fraction of fluid in that cell that is located behind the front. In this discussion, we identify the fluid behind the front as "black" and the fluid ahead as "white." Thus, a cell i,j has volume fraction $f_{i,j} = 1$ if it is entirely behind the front (black), a volume fraction $f_{i,j} = 0$ if it is entirely ahead of the front (white), and $f_{i,j}$ between 0 and 1 if it straddles the front. At any time step, an approximation to the front can be constructed from this array of volume fractions. This interface is advanced under the given velocity field by updating the fractions in the mesh cells in the following manner: The motion of the front is split into a sweep in the x direction with velocity v_x, followed by a sweep in the z direction with velocity v_z, where the

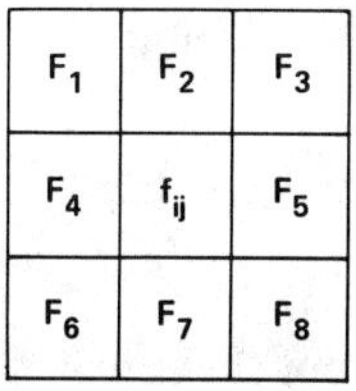

NUMERICAL METHODS FOR DISCONTINUOUS FLOWS 169

F_1	F_2	F_3
F_4	f_{ij}	F_5
F_6	F_7	F_8

FIG. 5.

velocity of the front is assumed to be (v_x, v_z). For each of these one-dimensional problems, an interface that represents an approximation to the front is drawn in each cell for which $0 < f_{i,j} < 1$. The orientation of the interface depends on the value of $f_{i,j}$ in both the cell and its neighbors. The "black" fluid is then advected in the $x(z)$ direction with velocity $v_x(v_z)$, and the new $f_{i,j}$'s are created, approximating the front advanced a distance $v_x \Delta t (v_z \Delta t)$, where Δt is the time step.

The original algorithm used line segments parallel to either the x- or z-axis to construct the local interface required for the one-dimensional sweeps. Numerous improvements have been made since. Trapezoids were added to the list of possible interface shapes, as well as thin slices of fluid to accommodate fingering. Since interchanging the order of the sweeps produced different results, the question of symmetry arose. One possible solution, of alternating the order of the sweeps, was found to be ineffective. Instead, both contributions were performed, and the new volume fraction in each cell was taken to be the maximum of the two results. This preserved symmetry by removing a bias towards the first sweep present in earlier calculations.

In applying this algorithm to porous flow problems, interfaces oblique to the grid directions were allowed in [25]. As an example of such an interface, consider a cell $f_{i,j}$ with neighbors $F_1, F_2, F_3, F_4, F_5, F_6, F_7$, and F_8, as in Fig. 5. We assume that the sweep is in the horizontal direction to the right, i.e., $v_x > 0$, and assume that, for example, $F_4 \neq 0$, $F_2 \neq 0$, $F_7 \neq 0$ and $F_5 = 0$. We are interested in establishing the location of the front within the center cell. From these volume fractions, we see that fluid lies on the top, bottom, and left, and thus assume that the front lies roughly parallel to the z-axis, as in Fig. 6. The slope of the interface

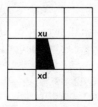

FIG. 6.

is determined from the ratio of the amount of fluid above to the amount of fluid below, namely,

$$\frac{x_u}{x_d} = \frac{F_2}{F_7},$$

and the requirement that the area of the trapezoid equal $f_{i,j}$, i.e.,

$$\tfrac{1}{2}h(x_u + x_d) = h^2 f_{i,j},$$

where x_u and x_d are the length of the sides of the trapezoids as in Fig. 6, and h is the size of the mesh. To advance the front under the time step Δt, we must move the sloped interface under the appropriate velocity field. Let $v_t(v_b)$ be the horizontal velocity at the front along the top (bottom) of the center cell. We translate the nodes of the trapezoid as follows

$$x_u^{n+1/2} = x_u^n + v_t \cdot \Delta t, \qquad x_d^{n+1/2} = x_d^n + v_b \cdot \Delta t,$$

where the superscript $n + \tfrac{1}{2}$ refers to the result after the first sweep with time step Δt. There are three cases, depending on whether or not the trapezoid has moved into the right cell. In Case 1 (Fig. 7a), the front remains in the center cell, and the updated value of $f_{i,j}$ is easily seen to be

$$f_{i,j}^{n+1/2} = \tfrac{1}{2}h(x_u^{n+1/2} + x_d^{n+1/2}).$$

In Case 2 (Fig. 7b), one leg of the trapezoid has moved to the cell on the right. Letting $\Delta f_{i,j}$ be the amount of fluid that has entered the cell on the right, the center cell is updated to

$$f_{i,j}^{n+1/2} = \tfrac{1}{2}h(x_u^{n+1/2} + x_d^{n+1/2}) - \Delta f_{i,j},$$

while the cell on the right becomes

$$f_{i+1,j}^{n+1/2} = f_{i+1,j}^n + \Delta f_{i,j}.$$

In Case 3, both legs of the trapezoid have moved into the right cell (Fig. 7c), and

$$f_{i,j}^{n+1/2} = 1,$$

while

$$f_{i+1,j}^{n+1/2} = f_{i+1,j}^n + \Delta f_{i,j}.$$

This concludes the one-dimensional sweep for the center cell. The full set of possible interfaces is shown in Fig. 8; the reader is referred to [25] and [32] for details.

Two points are worth mentioning. First, for any given cell, the orientation of the interface in the x sweep may be different from that constructed in the z sweep. Second, for the majority of cells, the value of $f_{i,j}$ and the neighboring $f_{i,j}$'s will be either all 0 or 1, implying that the front is not nearby. In these cases, no

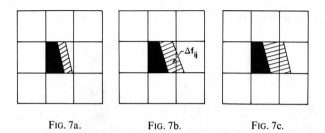

Fig. 7a.   Fig. 7b.   Fig. 7c.

calculations are required, since the value of $f_{i,j}$ will not change during the movement of the front. Judicious programming can avoid these situations altogether, reducing the calculation from $O(N^2)$ operations to $O(N)$ operations, where N^2 is the number of cells.

4.2. Algorithm for flow through porous media. We now summarize the algorithm presented in [25] for approximating the solution to problems of flow through porous media. The general idea is to use the above front tracking method to follow the discontinuity in s. At each time step, values for the pressure and velocity are obtained from s. This velocity field is used, together with the front tracking method, to move the line of discontinuity and update the saturation parameter.

We consider solving (2.1), (2.2) and (2.3) for the case $\gamma = 0$ (gravity effects absent) for porous flow in a domain Ω. For convenience, we absorb the porosity ϕ

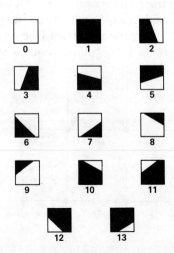

Fig. 8.

into the other variables to obtain

(4.1) $$\frac{\partial s}{\partial t} + \mathbf{q} \cdot \nabla f(s) = 0,$$

(4.2) $$\nabla \cdot \mathbf{q} = Q,$$

(4.3) $$\mathbf{q} = -\lambda(s)\nabla p.$$

Let s^n be the value of the saturation at time step $n\Delta t$. We assume that $s_{i,j}^n$ is known at the nodes of a square grid $\{i, j\}$ imposed on Ω. The pressures $p_{i,j}^n$ are taken at the same grid points, and the velocities q_x and q_z will be evaluated at the midpoints of the sides of the cell (see Fig. 9).

We shall describe the algorithm used to obtain $s_{i,j}^{n+1}$ from $s_{i,j}^n$. With $s_{i,j}^n$ known, first $p_{i,j}^n$ is calculated at the grid points. This is accomplished by substituting (4.3) into (4.2) to obtain an expression involving only p and $\lambda(s)$. A finite difference approximation to this expression is then solved, using the known values of $s_{i,j}^n$ to evaluate the necessary coefficient values for λ. Once the values of the pressure $p_{i,j}^n$ have been obtained, a finite difference approximation to (4.3) can be used to produce the velocity field $\mathbf{q}^n = (q_x^n, q_z^n)$.

To move to the next time step, the value of the saturation is updated according to (4.1). We use operator splitting to update s in two steps; a sweep in the x direction followed by a sweep in the z direction. The technique for solving the one-dimensional equation rests on our front tracking algorithm. Given the array of volume fractions $f_{i,j}^n$, we construct an approximation to the front, as described in the previous section. The Rankine–Hugoniot condition (§3) and the values of $\mathbf{q}^n$ and s^n provide the advection speed at the front. We compute the advection speed for all the cells of the front, and transport the black fluid to obtain the new $f_{i,j}^{n+1/2}$. Since we now know both the old and new positions of the front, we can compute the values of $s^{n+1/2}$ away from the front (the continuous parts) by any one of a variety of methods. In [25], both Godunov's method and a random

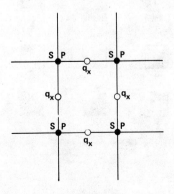

FIG. 9.

choice method are used to advance the solution. Given $s^{n+1/2}$, we complete the iteration by executing the sweep in the z direction, yielding the fully updated s^{n+1}.

4.3. Numerical results. In [25], numerical experiments were carried out for Ω the unit square, with Q corresponding to a source at $(0, 0)$ and a sink at $(1, 1)$, both of unit strength. We describe those experiments here. On $\partial\Omega$ the normal component of $\mathbf{q}$ and the normal derivative of s are both taken to be zero. Initially, the square is occupied entirely by "white" fluid to be displaced (i.e., $s = 0$), except at the point $(0, 0)$ at which $s = 1$. A 40×40 grid was used, with time step small enough so that during one time step, (i) the front can travel at most one cell length and (ii) the waves propagating from individual mesh-point discontinuities do not intersect. At $t = 0$, the square is filled with fluid to be displaced. A source is placed at $(0, 0)$ and a sink at $(1, 1)$, both of unit strength.

For the first test problem, the functions

$$f(s) = \frac{s^2}{s^2 + \alpha(1 - s)^2}, \qquad \lambda(s) = s^2 + \alpha(1 - s)^2$$

were used. This corresponds to a phase mobility proportional to the square of the saturation, and is representative of water flooding of a petroleum reservoir. The quantity α is the ratio of the viscosity of the wetting fluid to that of the nonwetting fluid (such as that of water to oil, respectively). Solutions were calculated for several values of α, corresponding to mobility ratios M at the front

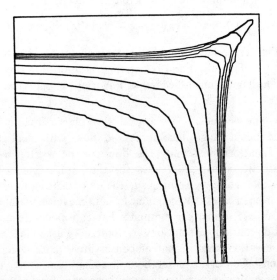

FIG. 10.

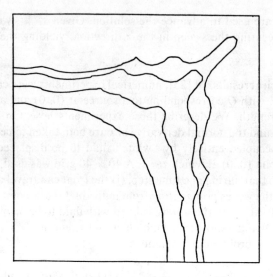

Fig. 11.

in both the stable ($M < 1$) and the unstable ($M > 1$) range. Figure 10 shows the results of a calculation with the above algorithm. With a value of $M = .845$, the front remains stable, as it should, and perturbations die out. In Figure 11, the same equations are solved with $M = 1.397$, a value in the unstable range. In this case, waves develop in the interface and fingering occurs.

The second test problem is one for miscible displacement. For this problem

$$f(s) = s, \quad \lambda(s) = s + M^{-1/4}(1 - s)^4$$

were used. In Fig. 12 we show the results of a calculation with $M = 2$, well into the unstable range. The initial and boundary conditions are the same as in the previous case. The front is unstable, and a "fingering" effect is clearly visible.

5. Godunov-type methods. For a large class of physically interesting cases of flow in porous media (e.g., incompressible flow with negligible capillary pressure), the equations describing the flow can be written as a system of nonlinear hyperbolic conservation laws for the saturations, with an elliptic equation for the total velocity. Solutions to the hyperbolic equations can develop discontinuities, even in cases in which none exist in the initial data. Consequently, it is necessary to use numerical methods that can calculate accurately both continuous and discontinuous solutions. A commonly used class of methods for hyperbolic conservation laws is that of conservative finite difference or finite elements methods. Conservative methods are ones for which the difference equations for the saturations are in discrete divergence form, guaranteeing that the total amount of each component of the fluid is conserved exactly. If the

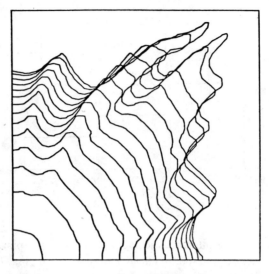

Fig. 12.

method is in discrete divergence form, then discontinuities are constrained to propagate, at least in some average sense, at the correct velocity.

One class of conservative finite difference techniques for calculating discontinuous solutions to systems of hyperbolic conservation laws was introduced by Godunov [19], [20] for gas dynamics. Godunov's method is a generalization to systems of nonlinear equations of the upwind differencing method for scalar advection equations. As such, it is generally too diffusive to represent discontinuities accurately. However, the higher order extensions of Godunov's method, first introduced by van Leer and then developed by a number of authors (for a review, see [22], [36]) have been demonstrated to be effective in calculating complicated time-dependent discontinuous solutions to the equations of gas dynamics in two space dimensions. Consequently, there is reason to believe that these methods will prove useful in calculating solutions to multidimensional problems arising in petroleum reservoir simulation.

We shall discuss the Godunov methods only for the case of one space variable. The one-dimensional form of these methods has been successfully used in multidimensional problems by means of operator splitting [5]. Multidimensional methods that do not require operator splitting are currently under investigation.

5.1. Scalar equations. We consider first Godunov's method for the scalar equation (3.1). We assume that at time t^n we know s_j^n, the average of s across each mesh interval $[(j - \frac{1}{2})\Delta z, (j + \frac{1}{2})\Delta z]$.

$$s_j^n = \frac{1}{\Delta z} \int_{(j-1/2)\Delta z}^{(j+1/2)\Delta z} s(z, t^n) \, dz.$$

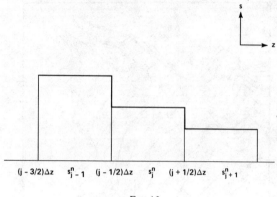

Fig. 13.

In Godunov's method, we interpret the averages s_j^n as giving piecewise constant interpolation functions of the solution in each mesh interval (Fig. 13)

(5.1) $\qquad s(z, t^n) = s_j^n, \qquad (j - \tfrac{1}{2})\Delta z < s < (j + \tfrac{1}{2})\Delta z.$

Since we know the solution to the Reimann problem, we can solve the initial value problem given by (5.1) exactly (Fig. 14), for a time Δt sufficiently small so that the waves from successive Riemann problems do not intersect. We denote the exact solution by $s_e^n(z, t)$. In order to obtain s_j^{n+1}, the average of the solution at the new time, we average $s_e^n(z, t)$ over the jth mesh interval (Fig. 15)

(5.2) $\qquad s_j^{n+1} = \dfrac{1}{\Delta z} \displaystyle\int_{(j-1/2)\Delta z}^{(j+1/2)\Delta z} s_e^n(z, t^n + \Delta t)\, dz.$

If the solution has complicated spatial structure, the evaluation of the integral in

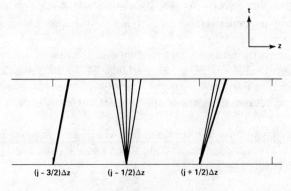

Fig. 14.

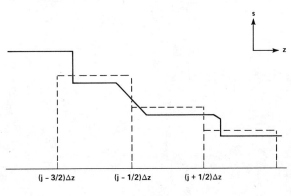

FIG. 15.

(5.2) can be difficult. By applying the divergence theorem to

$$\int_{(j-1/2)\Delta z}^{(j+1/2)\Delta z} \int_{t^n}^{t^{n+\Delta t}} \left(\frac{\partial s}{\partial t} + \frac{\partial \psi}{\partial z} \right) dt \, dz,$$

we obtain a difference formula for s_j^{n+1}.

(5.3) $\quad s_j^{n+1} - s_j^n = \frac{1}{\Delta z} \int_{t^n}^{t^n+\Delta t} (\psi(s_e^n((j - \tfrac{1}{2})\Delta z, t)) - \psi(s_e^n((j + \tfrac{1}{2})\Delta z, t))) \, dt.$

We observe that $s_e^n((j + \tfrac{1}{2})\Delta z, t) = s_{j+1/2}^{n+1/2}$, independent of t, where $s_{j+1/2}^{n+1/2}$ is obtained by evaluating the solution to the Riemann problem for (3.1) along the ray $(z - (j - \tfrac{1}{2})\Delta z)/(t - t^n) = 0$, with left and right states s_j^n, s_{j+1}^n. Thus we obtain

(5.4) $\quad\quad\quad\quad s_j^{n+1} = s_j^n + \frac{\Delta t}{\Delta z} (\psi(s_{j-1/2}^{n+1/2}) - \psi(s_{j+1/2}^{n+1/2})).$

The scheme is first order accurate, and is stable if $(\Delta t / \Delta z) \max_j |\psi'(s_j^n)| \leq 1$.

We now restrict our example further, by assuming $\psi'(s) \geq 0$, for all s. The solution to the Riemann problem at $(z - (j - \tfrac{1}{2})\Delta z)/(t - t^n) = 0$ is always the left state, implying that $s_{j+1/2}^{n+1/2} = s_j^n$ in (5.4). Thus Godunov's method in this situation reduces to upwind differencing, which is excessively dissipative. To obtain an algorithm with less dissipative error we replace the piecewise constant interpolation function (5.1) with one that is more accurate, and use the wave propagation properties of the equation to derive a difference scheme of the form (5.4).

The simplest such interpolation function we might use is a piecewise linear interpolation function (Fig. 16)

(5.5) $\quad s(z, t^n) = s_j^n + \delta s_j \dfrac{(z - j\Delta z)}{\Delta z}, \quad (j - \tfrac{1}{2})\Delta z < z < (j + \tfrac{1}{2})\Delta z,$

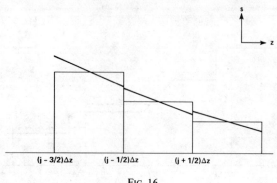

Fig. 16.

where $\delta s_j/\Delta z$ approximates $\partial s/\partial z|_{j\Delta z}$, subject to some constraints described below. Unlike the case for the piecewise constant interpolant, it is difficult to solve the piecewise linear problem analytically. Consequently, the difference scheme for the piecewise linear case is derived by approximating the time integrals in (5.3). If we approximate them using the midpoint rule, we obtain

$$s_j^{n+1} \approx s_j^n + \frac{\Delta t}{\Delta z}[\psi(s_e^n((j - \tfrac{1}{2})\Delta z, t^n + \tfrac{1}{2}\Delta t))$$
$$- \psi(s_e^n((j + \tfrac{1}{2})\Delta z, t^n + \tfrac{1}{2}\Delta t))].$$

We then approximate the value $s_e^n((j - \tfrac{1}{2})\Delta z, t^n + \tfrac{1}{2}\Delta t)$ by using the fact that solutions to (3.1) are constant along characteristics. If we approximate the characteristic through $((j + \tfrac{1}{2})\Delta z, t^n + \tfrac{1}{2}\Delta t)$ by the straight line (Fig. 17)

$$z(t) = (j + \tfrac{1}{2})\Delta z + (t - (t^n + \tfrac{1}{2}\Delta t))\psi'(s_j^n),$$

we obtain, using our interpolation function (5.5)

$$s_e^n((j + \tfrac{1}{2})\Delta z, t^n + \tfrac{1}{2}\Delta t) \approx s((j + \tfrac{1}{2})\Delta z - \tfrac{1}{2}\Delta t\, \psi'(s_j^n), t^n)$$
$$= s_j^n + \frac{1}{2}\left(1 - \frac{\Delta t}{\Delta z}\psi'(s_j^n)\right)\delta s_j.$$

Collecting the approximations, we obtain the difference scheme

(5.6)
$$s_j^{n+1} = s_j^n + \frac{\Delta t}{\Delta z}(\psi(s_{j-1/2}^{n+1/2}) - \psi(s_{j+1/2}^{n+1/2})),$$

$$s_{j+1/2}^{n+1/2} = s_j^n + \frac{1}{2}\left(1 - \frac{\Delta t}{\Delta z}\psi'(s_j^n)\right)\delta s_j.$$

To complete the specification of the scheme, we need to define δs_j. It is defined in two steps. First, we calculate a preliminary value $\widetilde{\delta s}_j$ using a central difference formula, e.g., $\tfrac{1}{2}\widetilde{\delta s}_j = s_{j+1} - s_{j-1}$ (for other examples, see [11], [35]). We then

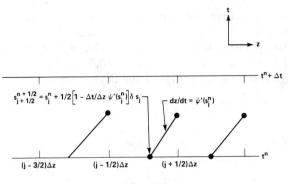

Fig. 17.

obtain our final value for δs_j by constraining $\widetilde{\delta s}_j$ to be within certain bounds. The purpose of the constraints is to prevent overshoots and undershoots at discontinuities. For example, the dashed profile in Fig. 18 represents a piecewise linear interpolant for which the left extrapolated value is out of the range defined by the adjacent zone averages. In that case, the slope is reduced so that the extrapolated value lies just within range, as represented by the dotted line. Also, if a zone average is a local extremum, then δs_j is set to zero. Expressed quantitatively, these constraints are given by

$$(5.7) \quad \delta s_j = \begin{cases} \min(|\widetilde{\delta s}_j|, 2|s_{j+1} - s_j|, 2|s_j - s_{j-1}|) \, \mathrm{sign}\,(s_{j+1} - s_{j-1}) \\ \qquad\qquad\qquad\qquad\qquad \text{if } (s_{j+1} - s_j)(s_j - s_{j-1}) > 0, \\ 0 \qquad\qquad\qquad\qquad\qquad \text{otherwise.} \end{cases}$$

In smooth parts of the solution, these inequalities are already satisfied by $\widetilde{\delta s}_j$, so that $\delta s_j = \widetilde{\delta s}_j$. With this choice of δs_j, it is not difficult to show that, if the solution is smooth, the scheme is formally second-order accurate.

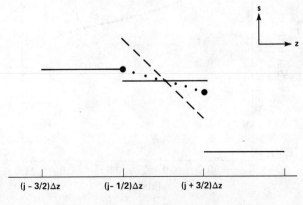

Fig. 18.

A numerical comparison between the first-order and second-order Godunov methods is depicted in Figs. 19 and 20. For these problems $s(z, t) = 1$ for $0 \leq z \leq 0.2$ and $s(z, 0) = 0$ for $0.2 < z \leq 1$. The dashed curves represent the solution of (3.1) with $\psi(s) = f(s) = s^2/(s^2 + 0.5(1 - s)^2)$ and $\Delta z = 0.02$. The first order method is depicted in Fig. 19 and the second order one in Fig. 20. The

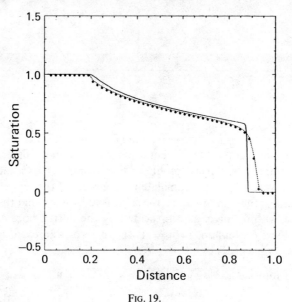

FIG. 19.

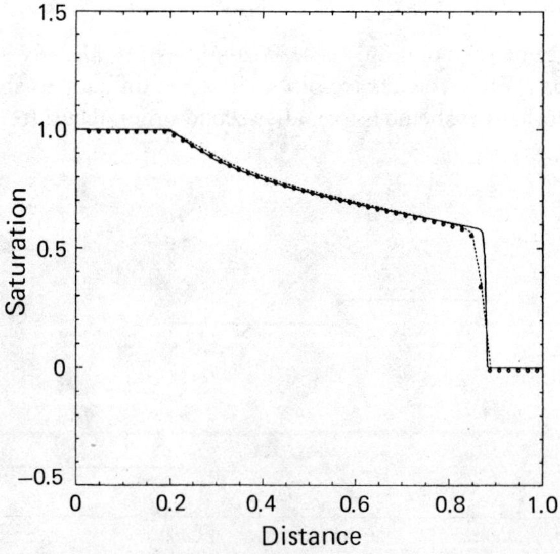

FIG. 20.

plotting routine indicates the data points by placing circles below them, more or less tangent to the interpolating curve. The solid lines in the figures represent the solution for $\Delta z = 0.0025$ using the second order method, which is essentially the exact solution for this case (data points are not indicated). The time step Δt was taken to be $0.1\,\Delta z$ for all cases, which corresponds to a CFL number of about 0.2. The improvement in the solution using the second-order over the first-order method is easily seen for this problem.

5.2. Systems of equations. We wish to extend the techniques described above to the case of the initial value problem for systems of hyperbolic conservation laws in one space variable

(5.8)
$$\frac{\partial U}{\partial t} + \frac{\partial F(U)}{\partial z} = 0,$$
$$U(z, t) = U{:}\mathbb{R} \times [0, T] \to \mathbb{R}^N,$$
$$F(U) = F{:}\mathbb{R}^N \to \mathbb{R}^N, \qquad U(z, 0) = U_0(z) \text{ given.}$$

In porous flow problems, U might be a vector of saturations, and $F(U)$ the vector of associated fractional flows. The system is assumed to be hyperbolic, i.e., the matrix $\nabla_U F = A(U)$ has N real eigenvalues $\lambda_1(U) < \cdots < \lambda_N(U)$ with corresponding left and right eigenvectors $(l_1, r_1), \cdots, (l_N, r_N)$. These eigenvectors are linearly independent and biorthogonal, i.e., $l_i \cdot r_j = 0$ if $i \neq j$. If one expands an arbitrary vector W in terms of the r_k's, then it follows from the biorthogonality property that the expansion coefficients are given by

$$W = \sum_{k=1}^{N} \alpha_k r_k, \qquad \alpha_k = l_k \cdot W.$$

These eigenvectors and eigenvalues are used to describe the infinitesimal wave propagation properties of the system (5.8). The characteristic curves of (5.8) are curves in (z, t) satisfying the ordinary differential equations

(5.9)
$$\frac{dz_k}{d\sigma_k} = \lambda_k(U(z_k(\sigma_k), t_k(\sigma_k))), \qquad \frac{dt_k}{d\sigma_k} = 1, \qquad k = 1, \cdots, N.$$

In regions where the solution is not discontinuous, a solution to (5.1) satisfies for each k an ordinary differential equation along the kth characteristic curve

(5.10)
$$l_k \cdot \frac{d}{d\sigma_k}(U(z_k(\sigma_k), t_k(\sigma_k))) = 0, \qquad k = 1, \cdots, N.$$

Derivatives of the solution are transported along characteristics, the component of the derivative transported along the kth characteristic being proportional to r_k.

As in the scalar case, we assume that U_j^n, the average value of the solution across a mesh interval, is known at time t^n

$$U_j^n = \frac{1}{\Delta z} \int_{(j-1/2)\Delta z}^{(j+1/2)\Delta z} U(z, t^n) dz.$$

We then wish to calculate U_j^{n+1}, the average value of the solution at time $t^n + \Delta t$. As in the scalar case above, we do so in the following three steps: (i) interpolate piecewise linear approximations to the solution at time t^n; (ii) use the wave propagation properties of the solution, in the form of Riemann problems and characteristic equations to find approximate values $U_{j+1/2}^{n+1/2}$ to the solution at $((j + \frac{1}{2})\Delta z, t^n + \frac{1}{2}\Delta t)$; and (iii) perform a conservative finite difference step to find U_j^{n+1}, of the form

(5.11) $$U_j^{n+1} = U_j^n + \frac{\Delta t}{\Delta z}(F(U_{j-1/2}^{n+1/2}) - F(U_{j+1/2}^{n+1/2})).$$

Steps (i) and (ii) are at the heart of the method. In these two steps, we generalize the algorithm given above for scalar equations by applying it one mode of wave propagation at a time, using the characteristic equations (5.9), (5.10). Since the characteristic form of the equations breaks down at shocks, care is required to guarantee that the algorithm reverts to something well-behaved near shocks.

The interpolation step is a straightforward generalization of what was done in the scalar case, except that the constraints are imposed in characteristic variables. We first calculate the preliminary value for the slope $\delta \tilde{U}_j = \frac{1}{2}(U_{j+1} - U_{j-1})$. We then modify the slope, using Harten's monotonicity algorithm for characteristic variables [21]. We expand

$$2(U_{j+1} - U_j) = \sum_{k=1}^{N} \bar{\alpha}_k^R r_k(U_j),$$

$$2(U_j - U_{j-1}) = \sum_{k=1}^{N} \bar{\alpha}_k^L r_k(U_j),$$

$$\delta \tilde{U}_j = \sum_{k=1}^{N} \tilde{\alpha}_k r_k(U_j).$$

We then define

$$\alpha_k = \begin{cases} \min(|\bar{\alpha}_k^L|, |\bar{\alpha}_k^R|, |\tilde{\alpha}_k|) \times \text{sign}(\tilde{\alpha}_k) & \text{if } \bar{\alpha}_k^L \cdot \bar{\alpha}_k^R > 0, \\ 0 & \text{otherwise.} \end{cases}$$

The constrained slope is then given by

$$\delta U_j = \sum \alpha_k r_k(U_j).$$

Given these interpolated profiles, we can now calculate $U_{j+1/2}^{n+1/2}$. The difficulty in calculating $U_{j+1/2}^{n+1/2}$ is that we want two rather different answers in two different limits.[1] In smooth parts of the solution we want $U_{j+1/2}^{n+1/2}$ to satisfy a finite difference approximation to the characteristic form of the equations (5.9), (5.10), i.e., we want it to satisfy the N linear equations

$$l_k \cdot (U_{j+1/2}^{n+1/2} - U_{j+1/2,k}), \qquad k = 1, \ldots, N,$$

where $U_{j+1/2,k}$ is the value of the solution at the base of the kth characteristic and is given by

$$U_{j+1/2,k} = \begin{cases} U_j^n + \dfrac{1}{2}\left(1 - \dfrac{\Delta t}{\Delta z}\lambda_k\right)\delta U_j & \text{if } \lambda_k > 0, \\ U_{j+1}^n - \dfrac{1}{2}\left(1 + \dfrac{\Delta t}{\Delta z}\lambda_k\right)\delta U_j & \text{if } \lambda_k < 0. \end{cases}$$

In the case for which $(j + \frac{1}{2})\Delta z$ is inside a discontinuity, our interpolated profile may look like the profile in Fig. 21, due to the constraints on δU. In that case, we want our solution $U_{j+1/2}^{n+1/2}$ to be given by the Riemann problem with left and right states corresponding to the jump at $(j + \frac{1}{2})\Delta z$, plus some perturbation representing the slopes on either side. One algorithm for $U_{j+1/2}^{n+1/2}$ that has the appropriate behavior is given in [13] for gas dynamics, but extends easily to this more general context.

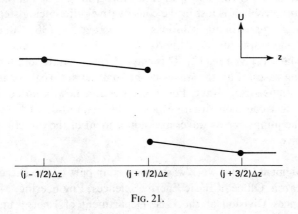

FIG. 21.

We take $U_{j+1/2}^{n+1/2}$ to be the solution to the Riemann problem with left and right states $U_{j+1/2,L}, U_{j+1/2,R}$, given by

(5.12)
$$U_{j+1/2,L} = U_{j+1/2,L}^n - A_{j+1/2,L}\,\delta U_j,$$
$$U_{j+1/2,R} = U_{j+1/2,R}^n - A_{j+1/2,R}\,\delta U_{j+1}.$$

---

[1] In the scalar results presented above, we avoided this problem by assuming $\psi'(s) \geq 0$.

Here $U^n_{j+1/2,L} = U^n_j + \tfrac{1}{2}\delta U_j$, $U^n_{j+1/2,R} = U^n_{j+1} - \tfrac{1}{2}\delta U_{j+1}$, and the operators $A_{j+1/2,L}$, $A_{j+1/2,R}$ are sums of the characteristic projection operators

$$A_{j+1/2,L} W = \sum_{k:\lambda_k(U^n_j)>0} \frac{1}{2}\lambda_k(U^n_j) \frac{\Delta t}{\Delta z}(l_k(U^n_j) \cdot W) r_k(U^n_j),$$

$$A_{j+1/2,R} W = \sum_{k:\lambda_k(U^n_{j+1})<0} \frac{1}{2}\lambda_k(U^n_{j+1}) \frac{\Delta t}{\Delta z}(l_k(U^n_{j+1}) \cdot W) r_k(U^n_{j+1}).$$

The vector $U_{j+1/2,L}$ ($U_{j+1/2,R}$) is equal to the left (right) limiting value of U at $((j+\tfrac{1}{2})\Delta z, t^n)$ plus the amount of wave of each family contained in $\delta U_j (\delta U_{j+1})$ that can reach $(j+\tfrac{1}{2})\Delta z$ from the left (right) between time t^n and $t^n + \tfrac{1}{2}\Delta t$. If $|\delta U_j|, |\delta U_{j+1}| \ll |U_j - U_{j+1}|$, then it is clear from (5.12) that $U^{n+1/2}_{j+1/2}$ is given by a small perturbation of the solution to the Riemann problem. If the solution is smooth, then it is not difficult to show that $U^{n+1/2}_{j+1/2}$ approximates, to second order, a solution to the characteristic equations (5.9), (5.10). This follows from the fact that, for weak waves, the solution to the Riemann problem reduces to transport along characteristics.

Throughout this discussion, we have assumed that the Riemann problem for (5.1) could be easily solved. In fact, this has been shown to be the case for only a few of the systems of equations arising in multiphase flow in porous media [23], [34], [35]. However, it is possible to introduce approximations into the solution of the Riemann problem without loss of accuracy, since much of the information in the Riemann problem is lost in the conservative differencing step. In particular, a class of approximate solutions is proposed in [30], and an explicit constructive algorithm for such approximate solutions for general systems of conservation laws is given in [12]. This class of approximate solutions is accurate in two limiting cases: if all the waves in the solution are weak, or if the solution consists of a single strong wave. For incompressible flow in porous media, these are the two most common situations since the magnitude and the direction in U-space of the jumps across waves are independent of the direction of propagation of the wave.

Acknowledgments. This work was supported in part by the Director, Office of Energy Research, Office of Basic Energy Sciences, Engineering, Mathematical, and Geosciences Division of the U.S. Department of Energy under contract DE-AC03-76SF00098.

REFERENCES

[1] N. ALBRIGHT, C. ANDERSON, AND P. CONCUS, *The random choice method for calculating fluid displacement in a porous medium*, Boundary and Interior Layers—Computational and Asymptotic Methods, J.J.H. Miller, ed., Boole Press, Dublin, 1980, pp. 3–13.

[2] N. ALBRIGHT AND P. CONCUS, *On calculating flows with sharp fronts in a porous medium,* Fluid Mechanics in Energy Conservation, J. D. Buckmaster, ed., Society for Industrial and Applied Mathematics, Philadelphia, 1980, pp. 172–184.

[3] N. ALBRIGHT, P. CONCUS, AND W. PROSKUROWSKI, *Numerical solution of the multidimensional Buckley–Leverett equation by a sampling method,* SPE 7681, 5th SPE Symposium on Reservoir Simulation, Denver, 1979.

[4] C. ANDERSON AND P. CONCUS, *A stochastic method for modeling fluid displacement in petroleum reservoirs,* in Analysis and Optimization of Systems, A. Bensoussan and J. L. Lions, eds., Lecture Notes in Control and Information Sciences 28, Springer-Verlag, Berlin, pp. 827–841.

[5] J. B. BELL, P. COLELLA, P. CONCUS, AND H. GLAZ, *Higher order Godunov methods for the Buckley-Leverett equation,* presented at SIAM 30th Anniversary Meeting, Stanford, CA, 1982.

[6] A. J. CHORIN, *Random choice solution of hyperbolic systems,* J. Comp. Phys., 22 (1976), pp. 517–533.

[7] ———, *Random choice methods with applicatons to reacting gas flow,* J. Comp. Phys., 25 (1977), pp. 253–272.

[8] ———, *Flame advection and propagation algorithms,* J. Comp. Phys., 35 (1980), pp. 1–11.

[9] ———, *Instability of fronts in a porous medium.* Report LBL-15893, Lawrence Berkeley Lab., Univ. California, Berkeley 1983; Comm. Math. Phys., to appear.

[10] P. COLELLA, *Glimm's method for gas dynamics,* SIAM J. Sci. Statist. Comput., 3 (1982) pp. 76–110.

[11] ———, *A direct Eulerian MUSCL scheme for gas dynamics,* Report LBL-14014, Lawrence Berkeley Lab., Univ. California, Berkeley 1982; SIAM J. Sci. Stat. Comput., to appear.

[12] ———, *Approximate solution of the Riemann problem for real gases,* Report LBL-14442, Lawrence Berkeley Lab., Univ. California, Berkeley 1983; J. Comp. Phys., to appear.

[13] P. COLELLA AND P. R. WOODWARD, *The piecewise-parabolic method* (PPM) *for gasdynamical simulations,* Report LBL-14661, Lawrence Berkeley Lab., Univ. California, Berkeley, 1983; J. Comp. Phys., to appear.

[14] P. CONCUS, *Calculation of shocks in oil reservoir modeling and porous flow,* in Numerical Methods for Fluid Dynamics, K. W. Morton and M. J. Baines, eds., Academic Press, New York, 1982, pp. 165–178.

[15] P. CONCUS AND W. PROSKUROWSKI, *Numerical solution of a nonlinear hyperbolic equation by the random choice method,* J. Comp. Phys., 30 (1979), pp. 153–166.

[16] M. CRANDALL AND A. MAJDA, *The method of fractional steps for conservation laws,* Numer. Math., 34 (1980), pp. 285–314.

[17] J. GLIMM, *Solutions in the large for nonlinear hyperbolic systems of equations,* Comm. Pure Appl. Math., 18(1965), pp. 697–715.

[18] J. GLIMM, D. MARCHESIN AND O. MCBRYAN, *The Buckley–Leverett equation: theory, computation and application,* Proc. Third Meeting of the International Society for the Interaction of Mechanics and Mathematics, Edinburgh, 1979; Trends in Applications of Pure Mathematics in Mechanics, R. J. Knops, ed., Edinburgh, 1979.

[19] S. K. GODUNOV, *Finite difference methods for numerical computation of discontinuous solutions of the equations of fluid dynamics.* Mat. Sbornik, 47 (1959), pp. 271–306. (In Russian.)

[20] S. K. GODUNOV, A. V. ZABRODYN, AND G. P. PROKOPOV, *A computational scheme for two-dimensional nonstationary problems of gas dynamics and calculations of the flow from a shock wave approaching a stationary state,* USSR Comput. Math. Math. Phys., 1 (1961), pp. 1187–1218.

[21] A. HARTEN, *On second order accurate Godunov-type schemes,* preprint, 1982.

[22] A. HARTEN, P. D. LAX, AND B. VAN LEER, *On upstream differencing and Godunov-type schemes for hyperbolic conservation laws,* SIAM Rev., 25 (1983), pp. 35–62.

[23] E. ISAACSON, *Global solution of a Riemann problem for a non-strictly hyperbolic system of conservation laws arising in enhanced oil recovery,* preprint, Math. Dept., Rockefeller Univ., New York, 1982; J. Comp. Phys., to appear.

[24] P. D. LAX, *Hyperbolic Systems of Conservation Laws and the Mathematical Theory of Shock Waves,* CBMS Regional Conference Series in Applied Mathematics 11, Society for Industrial and Applied Mathematics, Philadelphia, 1973.

[25] P. LOTSTEDT, *A front tracking method applied to Burgers' equation and two-phase porous flow,* J. Comp. Phys., 47 (1982), pp. 211–228.

[26] W. F. NOH AND P. WOODWARD, SLIC (*Simple Line Interface Calculation*), Proc. 5th Intl. Conference on Numerical Methods in Fluid Dynamics, A. I. van de Vooren and P. J. Zandbergen, eds., Lecture Notes in Physics 59, Springer-Verlag, Berlin, 1976, pp. 330–340.

[27] O. A. OLEINIK, *Uniqueness and stability of the generalized solution of the Cauchy problem for a quasilinear equation,* AMS Translat., Ser. 2, 33 (1963), pp. 285–290.

[28] D. W. PEACEMAN, *Fundamentals of Numerical Reservoir Simulation,* Elsevier, Amsterdam, 1977.

[29] W. PROSKUROWSKI, *A note on solving the Buckley–Leverett equation in the presence of gravity,* J. Comp. Phys., 41 (1981), pp. 136–141.

[30] P. ROE, *Approximate Riemann solvers, parameter vectors, and difference schemes,* J. Comp. Phys., 43 (1981), pp. 357–372.

[31] J. A. SETHIAN, *An analysis of flame propagation,* Report LBL-14125, Lawrence Berkeley Lab., Univ. California, Berkeley, 1982.

[32] ———, *Turbulent combustion in open and closed vessels,* Report LBL-15744, Lawrence Berkeley Lab., Univ. California, Berkeley, 1983; J. Comp. Phys. (1984), to appear.

[33] B. TEMPLE, *Global solution for the Cauchy problem for a class of* 2×2 *nonstrictly hyperbolic conservation laws,* Adv. Appl. Math, 3 (1983), pp. 335–375.

[34] ———, *Systems of conservation laws with coinciding shock and rarefaction curves,* preprint; Proc. Amer. Math. Soc, to appear.

[35] B. VAN LEER, *Towards the ultimate conservative difference scheme. V. A second-order sequel to Godunov's method,* J. Comp. Phys., 32 (1979), pp. 101–136.

[36] P. R. WOODWARD AND P. COLELLA, *The numerical simulation of two-dimensional fluid flow with strong shocks,* J. Comp. Phys, to appear.